CORE FOCUS ON

RATIOS, RATES & STATISTICS

SMC curriculum

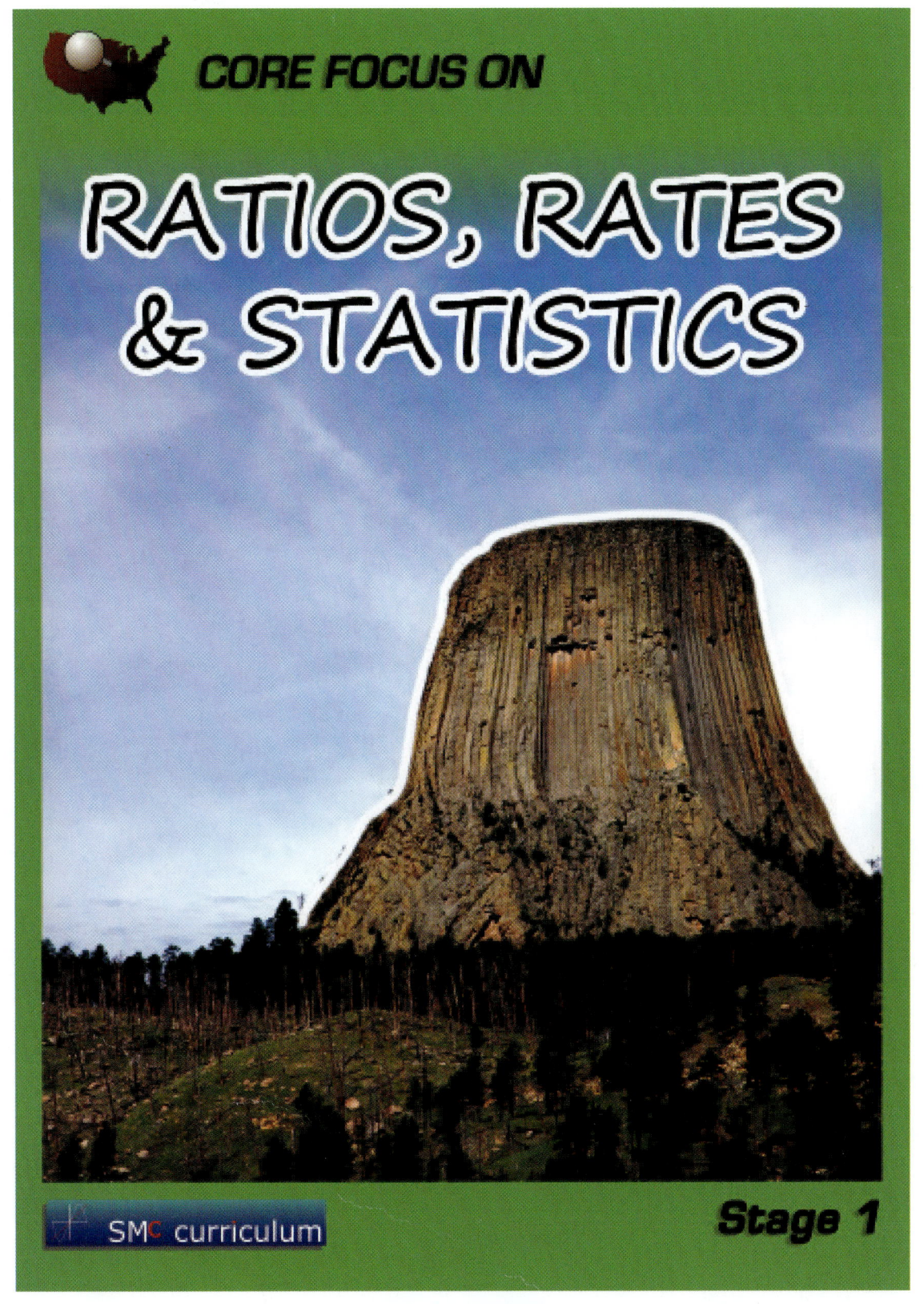

Stage 1

AUTHORS

Shannon McCaw

Beth Armstrong • Matt McCaw • Sarah Schuhl • Michelle Terry • Scott Valway

COVER PHOTOGRAPH

Devil's Tower National Monument
The first National Monument, Devil's Tower is a geologic wonder in the plains of northeastern Wyoming. Designated a monument in 1906, Devil's Tower is a popular place for rock climbers and visitors from around the world.
©iStockphoto.com/Jan Zoetekouw

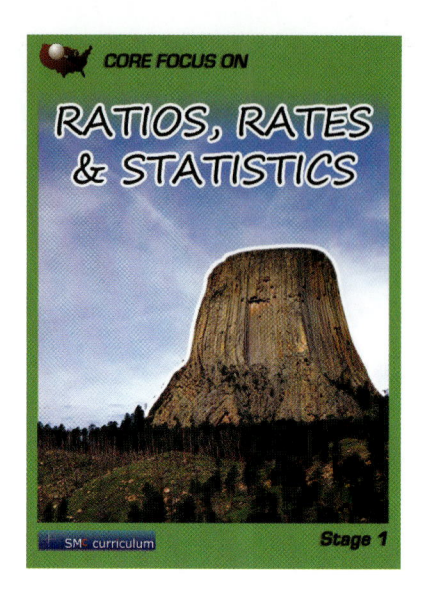

Copyright ©2014 by SMc Curriculum, LLC. All rights reserved. Printed in the United States. This publication is protected by copyright. No part of this publication should be reproduced or transmitted in any form or by any means without prior written consent of the publisher. This includes, but is not limited to, electronic reproduction, storage in a retrieval system, photocopying, recording or broadcasting for distance learning. For information regarding permission(s), write to: Permissions Department.

ISBN: 978-1-938801-71-6

1 2 3 4 5 6 7 8 9 10

ABOUT THE AUTHORS

From left to right: Beth Armstrong, Matt McCaw, Shannon McCaw, Scott Valway, Michelle Terry, Sarah Schuhl

SERIES AUTHOR

Shannon McCaw has been a classroom teacher in the Newberg and Parkrose School Districts in Oregon. She has been trained in Professional Learning Communities, Differentiated Instruction and Critical Friends. Shannon currently works as a consultant with math teachers from over 100 districts around Oregon. Shannon's areas of expertise include the Common Core State Standards, curriculum alignment, assessment best practices and instructional strategies. She has a degree in Mathematics from George Fox University and a Masters of Arts in Secondary Math Education from Colorado College.

CONTRIBUTING AUTHORS & EDITORS

Beth Armstrong has been a classroom teacher in the Beaverton School District in Oregon. She has received training in Talented and Gifted Instruction. She has a Masters in Curriculum and Instruction from Washington State University.

Matt McCaw has been a classroom teacher, math/science TOSA and special education case-manager in several Oregon school districts. Matt has most recently worked as a curriculum developer and math coach for grades 6-8. He is trained in Differentiated Instruction, Professional Learning Communities, Critical Friends Groups and Understanding Poverty. Matt has a Masters of Special Education from Western Oregon University.

Sarah Schuhl has been a classroom teacher, secondary math instructional coach and district-wide K-12 math instructional specialist, most recently in the Centennial School District in Oregon. Sarah currently works as a Solution Tree associate and an educational consultant supporting and challenging teachers in the areas of math instruction and alignment to the Common Core State Standards, common assessments for all subjects and grade levels and professional learning communities. From 2010–2013, Sarah served as a member and chair of the National Council of Teachers of Mathematics editorial panel for their Mathematics Teacher journal. Sarah earned a Masters of Science in Teaching Mathematics from Portland State University.

Michelle Terry has been a classroom teacher in the Estacada and Newberg School Districts in Oregon. Michelle has received training in Professional Learning Communities, Critical Friends, ELL Instructional Strategies, Proficiency-Based Grading and Lesson Design, Power Strategies for Effective Teaching, and Classroom Love and Logic. Michelle has an Interdisciplinary Masters from Western Oregon University. She currently teaches mathematics at Newberg High School.

Scott Valway has been a classroom teacher in the Tigard-Tualatin, Newberg and Parkrose School Districts in Oregon. Scott has been trained in Differentiated Instruction, Professional Learning Communities, Critical Friends, Discovering Algebra, Pre-Advanced Placement, Assessment Writing and Credit by Proficiency. Scott has a Masters of Science in Teaching from Oregon State University. He currently teaches math at Parkrose High School.

COMMON CORE STATE STANDARDS
Grade 6 Overview

The complete set of Common Core State Standards can be found at http://www.corestandards.org/. This book focuses on the highlighted Common Core State Standards shown below.

Ratios and Proportional Relationships

- ==Understand ratio concepts and use ratio reasoning to solve problems.==

The Number System

- Apply and extend previous understanding of multiplication and division to divide fractions by fractions.

- Compute fluently with multi-digit numbers and find common factors and multiples.

- Apply and extend previous understandings of numbers to the system of rational numbers.

Expressions and Equations

- Apply and extend previous understandings of arithmetic to algebraic expressions.

- Reason about and solve one-variable equations and inequalities.

- Represent and analyze quantitative relationships between dependent and independent variables.

Geometry

- Solve real-world and mathematical problems involving area, surface area and volume.

Statistics and Probability

- ==Develop understanding of statistical variability.==

- ==Summarize and describe distributions.==

Mathematical Practices

1. Make sense of problems and persevere in solving them.

2. Reason abstractly and quantitatively.

3. Construct viable arguments and critique the reasoning of others.

4. Model with mathematics.

5. Use appropriate tools strategically.

6. Attend to precision.

7. Look for and make use of structure.

8. Look for and express regularity in repeated reasoning.

CORE FOCUS ON RATIOS, RATES & STATISTICS

CONTENTS IN BRIEF

How To Use Your Math Book	VIII
Block 1 Ratios and Conversions	1
Block 2 Rates	33
Block 3 Percents and Probability	71
Block 4 Statistics	115
Acknowledgements	167
English/Spanish Glossary	168
Selected Answers	203
Index	208
Problem-Solving	211
Symbols	212

CORE FOCUS ON RATIO, RATES & STATISTICS

BLOCK 1 ~ RATIOS AND CONVERSIONS

Lesson 1.1	Ratios	3
	Explore! Comparing Students	
Lesson 1.2	Geometric Sequences	9
	Explore! Number Patterns	
Lesson 1.3	Customary Conversions	14
Lesson 1.4	Metric Conversions	18
Lesson 1.5	Perimeter and Area	22
	Explore! The Patio	
Review	Block 1 ~ Ratios and Conversions	28

BLOCK 2 ~ RATES

Lesson 2.1	Fractions and Decimals	35
	Explore! Back and Forth	
Lesson 2.2	Repeating Decimals and Rounding	39
	Explore! Calculators and Fractions	
Lesson 2.3	Rates and Unit Rates	44
Lesson 2.4	Rate Problem Solving	48
	Explore! Match the Rates	
Lesson 2.5	Comparing Rates	54
	Explore! Shopping Sales	
Lesson 2.6	Motion Rates	60
Review	Block 2 ~ Rates	66

BLOCK 3 ~ PERCENTS AND PROBABILITY

LESSON 3.1	INTRODUCING PERCENTS	73
	EXPLORE! PERCENTS	
LESSON 3.2	PERCENTS, DECIMALS AND FRACTIONS	77
	EXPLORE! KIERAN'S ROOM	
LESSON 3.3	PERCENT OF A NUMBER	81
LESSON 3.4	PERCENT APPLICATIONS	85
	EXPLORE! AT THE RESTAURANT	
LESSON 3.5	INTRODUCTION TO PROBABILITY	90
LESSON 3.6	EXPERIMENTAL PROBABILITY	94
	EXPLORE! ROLLING A 3	
LESSON 3.7	THEORETICAL PROBABILITY	100
	EXPLORE! SUM OF TWO NUMBER CUBES	
LESSON 3.8	GEOMETRIC PROBABILITY	105
	EXPLORE! WHAT ARE MY CHANCES OF WINNING?	
REVIEW	BLOCK 3 ~ PERCENTS & PROBABILITY	109

BLOCK 4 ~ STATISTICS

LESSON 4.1	INTRODUCTION TO STATISTICS	117
	EXPLORE! A QUESTION OF STATISTICS	
LESSON 4.2	MEASURES OF CENTER	122
	EXPLORE! COUNTING PETS	
LESSON 4.3	DOT PLOTS	128
	EXPLORE! HOW TALL?	
LESSON 4.4	HISTOGRAMS	134
LESSON 4.5	BOX-AND-WHISKER PLOTS	141
LESSON 4.6	ANALYZING STATISTICS	148
	EXPLORE! WHAT'S THE "MEAN"ING?	
LESSON 4.7	MEAN ABSOLUTE DEVIATION	154
	EXPLORE! MERCURY'S RISING	
REVIEW	BLOCK 4 ~ STATISTICS	160

HOW TO USE YOUR MATH BOOK

Your math book has features that will help you be successful in this course. Use this guide to help you understand how to use this book.

LESSON TARGET

> Look in this box at the beginning of every lesson to know what you will be learning about in each lesson.

VOCABULARY

Each new vocabulary word is printed in **red**. The definition can be found with the word. You can also find the definition of the word in the glossary which is in the back of this book.

EXPLORE!

> Some lessons have **EXPLORE!** activities which allow you to discover mathematical concepts. Look for these activities in the Table of Contents and in lessons next to the purple line.

EXAMPLES

> Examples are useful because they remind you how to work through different types of problems. Look for the word **EXAMPLE** and the green line.

HELPFUL HINTS

> Helpful hints and important things to remember can be found in green callout boxes.

BLUE BOXES

> A blue box holds important information or a process that will be used in that lesson. Not every lesson has a blue box.

This calculator icon will appear in Lessons and Exercises where a calculator is needed. Your teacher may want you to use your calculator at other times, too. If you are unsure, make sure to ask if it is the right time to use it.

viii Core Focus on Ratios, Rates & Statistics

EXERCISES

The **EXERCISES** are a place for you to find practice problems to determine if you understand the lesson's target. You can find selected answers in the back of this book so you can check your progress.

REVIEW

The **REVIEW** provides a set of problems for you to practice concepts you have already learned in this book. The **REVIEW** follows the **EXERCISES** in each lesson. There is also a **REVIEW** section at the end of each Block.

TIC-TAC-TOE ACTIVITIES

Each Block has a Tic-Tac-Toe board at the beginning with activities that extend beyond the Common Core State Standards. The Tic-Tac-Toe activities described on the board can be found throughout each Block in yellow boxes.

CAREER FOCUS

At the end of each Block, you will find an autobiography of an individual. Each one describes what they like about their job and how math is used in their career.

CORE FOCUS ON MATH
STAGE 1

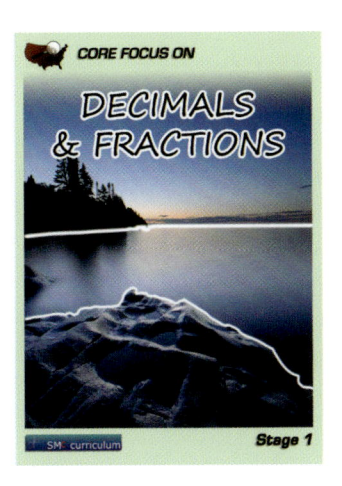

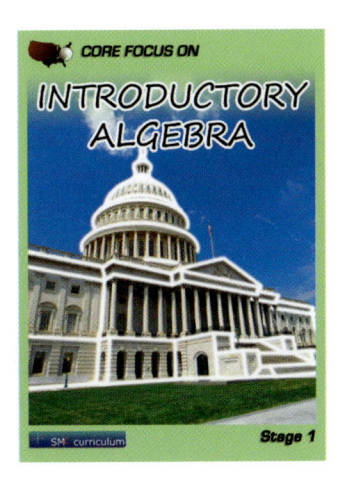

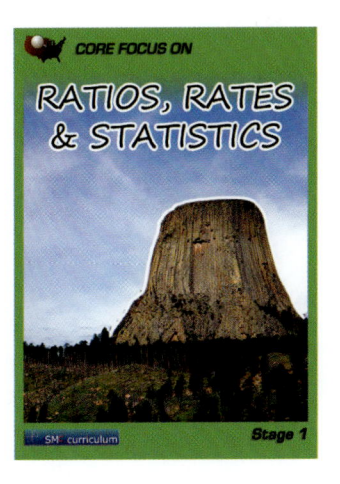

CORE FOCUS ON RATIOS, RATES & STATISTICS
BLOCK 1 ~ RATIOS AND CONVERSIONS

LESSON 1.1	RATIOS	3
	EXPLORE! COMPARING STUDENTS	
LESSON 1.2	GEOMETRIC SEQUENCES	9
	EXPLORE! NUMBER PATTERNS	
LESSON 1.3	CUSTOMARY CONVERSIONS	14
LESSON 1.4	METRIC CONVERSIONS	18
LESSON 1.5	PERIMETER AND AREA	22
	EXPLORE! THE PATIO	
REVIEW	BLOCK 1 ~ RATIOS AND CONVERSIONS	28

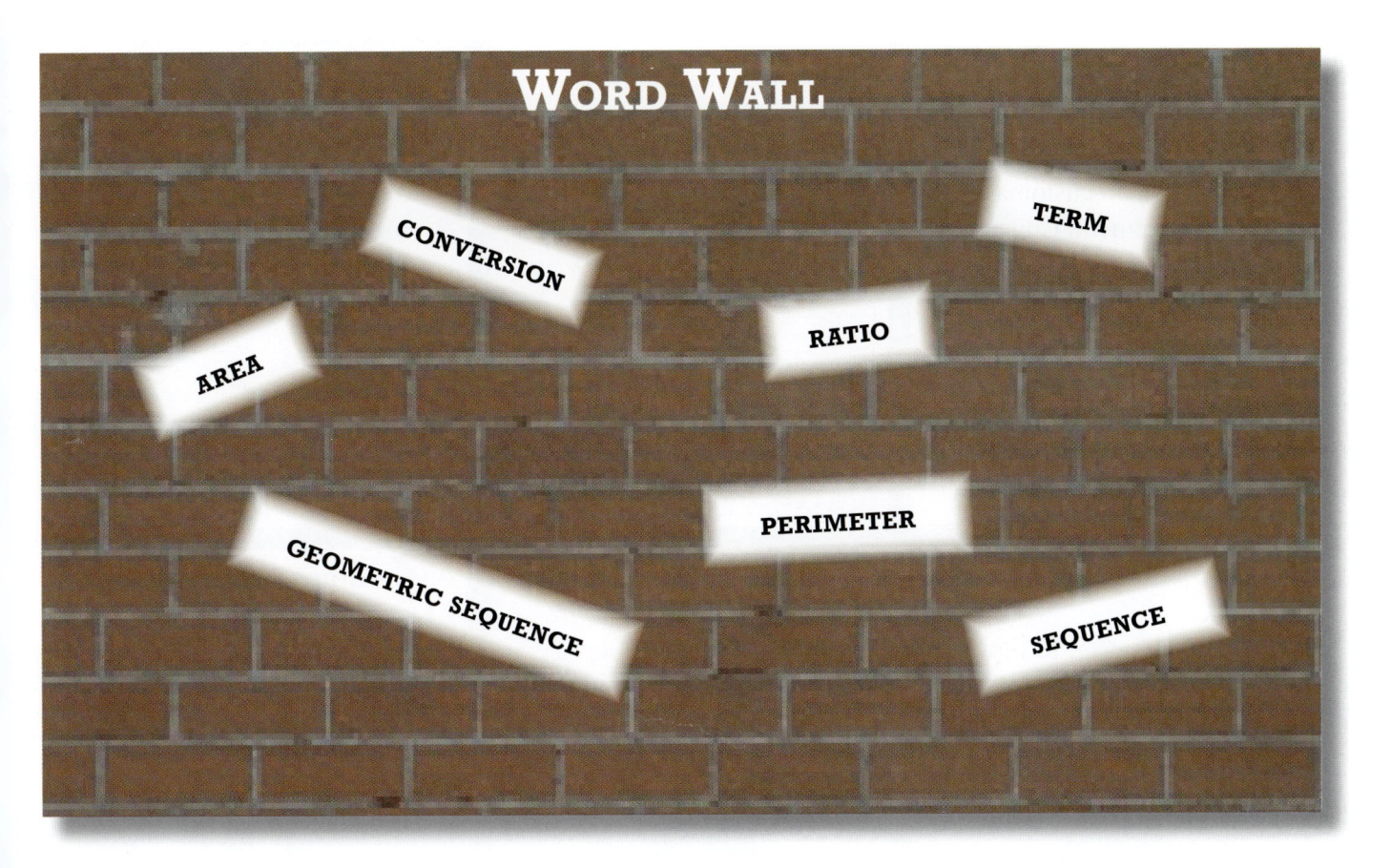

WORD WALL

CONVERSION
TERM
RATIO
AREA
PERIMETER
GEOMETRIC SEQUENCE
SEQUENCE

BLOCK 1 ~ RATIOS AND CONVERSIONS
TIC-TAC-TOE

PAY RAISE Choose a pay raise from two different options at your new job. *See page 27 for details.*	**CARPET COST** Measure three rooms in your house. Find the cost for two different colors of carpet. *See page 26 for details.*	**HEIGHT CONVERSIONS** Find the height in feet of ten objects. Convert the measurements to inches, yards and miles. *See page 17 for details.*
METRIC LENGTHS List countries that use metric and customary measurement. Convert landmarks in meters to *mm*, *cm* and *km*. *See page 21 for details.*	**STUDENT SURVEY** Survey students about something they like and determine ratios. *See page 21 for details.*	**GNAT STORY** Use a geometric sequence to find how many times a gnat can fly half the distance across a room. *See page 17 for details.*
SQUARE UNITS Make a poster to show how to convert units involving area. *See page 27 for details.*	**NUMBER SEQUENCES** Explore geometric sequences, arithmetic sequences and the Fibonacci sequence. *See page 13 for details.*	**BODY RATIOS** Measure your height, wingspan, foot and forearm to find ratios. *See page 8 for details.*

RATIOS

LESSON 1.1

🎯 Simplify and write ratios three ways.

Alexa counted the number of boys and girls in each of the 6th grade classes at her school. She recorded the information in the table below.

Class	Boys	Girls
Mr. Brown	15	15
Ms. Tran	12	20
Ms. Cruz	20	15
Mr. Graf	8	24

EXPLORE!   COMPARING STUDENTS

Step 1: Write a fraction comparing the number of boys to the number of girls for each teacher $\left(\frac{\text{number of boys}}{\text{number of girls}}\right)$. Write the fraction in simplest form. This fraction means for every ____ boys there are ____ girls.

Step 2: Find the total number of students in each teacher's class.

Write a fraction comparing the number of boys to the total number of students for each teacher $\left(\frac{\text{number of boys}}{\text{total number of students}}\right)$. Write the fraction in simplest form. This fraction means for every ____ boys there are ____ students.

Step 3: Write a fraction comparing the number of girls to the total number of students for each teacher $\left(\frac{\text{number of girls}}{\text{total number of students}}\right)$. Write the fraction in simplest form. This fraction means for every ____ girls there are ____ students.

Step 4: Count the number of boys and girls in your class.
  a. Find the fraction of boys to girls. This is the ratio of boys to girls.
  b. Find the ratio of boys to total number of students.
  c. Find the ratio of girls to total number of students.

Step 5: Mr. Jansen's math class has 3 boys for every 2 girls.
  a. Write a ratio of boys to girls in Mr. Jansen's class.
  b. There are 30 students in Mr. Jansen's class. How many are boys? Explain how you know your answer is correct.

A:B or A/B

A **ratio** uses division to compare two numbers. Each ratio shows a part to another part or a part to a whole. A ratio written as a fraction should be written in simplest form. There are two other ways to write ratios: using a colon or the word "to."

Lesson 1.1 ~ Ratios   3

WRITING RATIOS

A ratio comparing two numbers, 3 and 5, can be written in three ways:
- As a fraction → $\frac{3}{5}$
- Using a colon → 3 : 5
- Using the word "to" → 3 to 5

Ratios can be larger than 1. For example, 16 girls and 8 boys are in a class. The ratio of girls to boys is $\frac{16}{8}$ which simplifies to $\frac{2}{1}$. It can be written $\frac{2}{1}$, 2 : 1, or 2 to 1. Ratios larger than 1 should be written as simplified improper fractions.

EXAMPLE 1

Paul took a handful of jelly beans. He chose 4 blue, 2 green, 3 red and 3 yellow. Write a ratio using each of the three ways.
a. Compare blue jelly beans to red jelly beans.
b. Compare green jelly beans to the total number of jelly beans.
c. Compare red jelly beans to the total number of jelly beans.

SOLUTIONS

	Fraction	Using a Colon	Using "to"
a.	$\frac{\text{blue jelly beans}}{\text{red jelly beans}} \rightarrow \frac{4}{3}$	4 : 3	4 to 3
b. Total number of jelly beans: 4 + 2 + 3 + 3 = 12 $\frac{\text{green jelly beans}}{\text{total jelly beans}} \rightarrow \frac{2}{12} = \frac{1}{6}$		1 : 6	1 to 6
c. Total number of jelly beans: 4 + 2 + 3 + 3 = 12 $\frac{\text{red jelly beans}}{\text{total jelly beans}} \rightarrow \frac{3}{12} = \frac{1}{4}$		1 : 4	1 to 4

EXAMPLE 2

The ratio of boys to girls on a soccer team is 8 : 7. What is the ratio of boys to all players on the soccer team?

SOLUTION

The ratio 8 : 7 means there are 8 boys for every 7 girls.

For every 8 boys there are 15 (8 + 7) players.

This makes the ratio of boys to all players 8 : 15.

Leela is making chocolate chip cookies. The recipe states she needs 3 cups of sugar and 2 eggs. This means the ratio of sugar to eggs in the dough is 3 : 2 or $\frac{3}{2}$. If she doubles the recipe, she will need to double these ingredients. She will need 6 cups of sugar and 4 eggs. The ratio of sugar to eggs is now 6 : 4 or $\frac{6}{4}$. Notice that $\frac{6}{4} = \frac{3}{2}$ so these ratios are equivalent. The amount of each ingredient changed when Leela doubled the recipe, but the ratio of sugar to eggs did not change.

EXAMPLE 3

Compare the number of stars to the number of circles. If the ratio remains the same, how many stars will you have if you have 14 circles?

★ ★ ★ ★ ★   ● ●
★ ★ ★ ★ ★   ● ●

SOLUTION

Write the ratio of stars to circles as a fraction.

$$\frac{10}{4}$$

Simplify the ratio.

$$\frac{10}{4} = \frac{5}{2}$$

Find an equivalent ratio that has 14 circles.

$$\frac{5}{2} \; \overset{\times 7}{=} \; \frac{35}{14}$$

You will have 35 stars. The ratio 35 : 14 is equivalent to 10 : 4.

The different quantities used to make a ratio can also be written in a table and the points graphed.

EXAMPLE 4

Petra is filling jars with marbles. The jar sizes change, but the ratio of blue marbles to green marbles is always the same. The table below shows some of the jars and their quantity of blue marbles and green marbles.

blue	6	12	18	27	33
green	10	20	30	45	?

a. Find the ratio of blue marbles to green marbles in each jar.
b. Find the number of green marbles when the jar has 33 blue marbles.
c. Let the number of blue marbles be x and the number of green marbles be y. Plot the ordered pairs (x, y) for each set of values in the table. What do you notice about the points?

SOLUTIONS

a. Choose any pair of blue and green values and write the ratio as a fraction.

$$\frac{6}{10}$$

Simplify the ratio.

$$\frac{6}{10} = \frac{3}{5}$$

The ratio of blue marbles to green marbles is 3 : 5 or $\frac{3}{5}$.

b. Find an equivalent ratio to $\frac{3}{5}$ that has 33 blue marbles.

$$\frac{3}{5} \; \overset{\times 11}{=} \; \frac{33}{55}$$

There would be 55 green marbles when the jar has 33 blue marbles.

Lesson 1.1 ~ Ratios

EXAMPLE 4
SOLUTIONS
(CONTINUED)

c. Graph each of the ordered pairs (6, 10), (12, 20), (18, 30), (27, 45) and (33, 55).

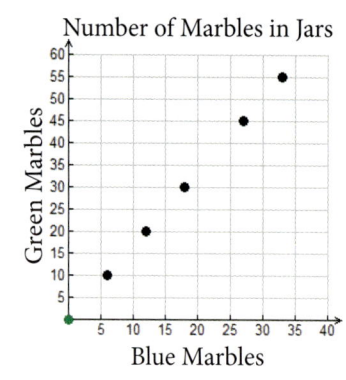

Number of Marbles in Jars

The points are all in a line with the origin (0, 0). Each point has a ratio of its *y*-value to its *x*-value of $\frac{5}{3}$.

EXERCISES

Write two different equivalent ratios for each ratio.

1. $\frac{2}{4}$
2. 5 to 20
3. 9 : 3
4. 4 : 12
5. 6 to 6
6. $\frac{30}{20}$

Simplify each ratio. Write each ratio as a fraction, with a colon and using the word "to".

7. $\frac{10}{15}$
8. 7 : 14
9. 8 to 2
10. $\frac{4}{6}$
11. 9 : 9
12. $\frac{100}{10}$

Write a ratio for each situation as a fraction in simplest form.

13. Seven girls and 8 boys are in leadership class.
    a. Write the ratio of girls to boys.
    b. Write the ratio of girls to total students in leadership class.
    c. Write the ratio of boys to total students in leadership class.

14. Three goldfish out of 12 goldfish in an aquarium ate their food.
    a. Write the ratio of goldfish that ate their food to the total number of goldfish in the aquarium.
    b. Write the ratio of goldfish that ate their food to those that did not.

15. Ten out of 20 sixth graders wore white tennis shoes in PE.
    a. Write the ratio of sixth graders who wore white tennis shoes to the total number of sixth graders in PE.
    b. Write the ratio of sixth graders who wore white tennis shoes to the number of sixth graders who did not wear white tennis shoes.

16. Eight black labs and 10 collies entered a dog competition.
    a. Write the ratio of black labs to collies.
    b. Write the ratio of black labs to the total number of dogs in the competition.

Lesson 1.1 ~ Ratios

17. Twelve out of 36 teachers live more than 5 miles from school.
    a. Write the ratio of teachers who live more than 5 miles from school to the total number of teachers.
    b. Write the ratio of teachers who live more than 5 miles from school to those who live closer than 5 miles from school.

18. An animal shelter had 7 cats, 8 dogs and 5 rabbits available for adoption. Find the ratio of rabbits to total animals at the shelter.

19. The Bobcats softball team had a win to loss ratio of 3 to 4.
    a. What is the meaning of the ratio 3 : 4?
    b. What is the ratio of wins to total games played?

20. One day Mrs. Garcia's class had a 5 : 1 ratio of students wearing shorts to long pants.
    a. What is the meaning of the ratio 5 : 1?
    b. What is the ratio of students wearing long pants to all students in Mrs. Garcia's class?

21. Julio asked several students what type of movie they preferred. He recorded the information in a chart.

Movie Type	Comedy	Drama	Action	Adventure	Cartoon
Number of Students	25	20	40	10	5

    a. Write a ratio representing the number of students who prefer comedy movies to the number of students who prefer action movies.
    b. Find the total number of students Julio interviewed.
    c. Write a ratio representing the number of students who preferred each type of movie to the total number of students interviewed. (You will have 5 different ratios.)

22. Matt and Shannon were trying to write ratios close to one. Matt wrote 3 : 4 and Shannon wrote $\frac{9}{10}$.
    a. Whose ratio is closest to 1? Explain how you know your answer is correct.
    b. Write a ratio closer to 1 than Matt's or Shannon's ratios.

23. Compare the number of squares to the number of triangles. If the ratio remains the same, how many squares will there be if there are 32 triangles?

24. Compare the number of cars to the number of trucks.
    a. If the ratio remains the same, how many trucks will there be when there are 28 cars? Use mathematics to justify your answer.
    b. Compare the number of cars to all vehicles. If there are 48 vehicles, how many cars are there? Use mathematics to justify your answer.

Each table is made of *x*- and *y*-coordinates that form equivalent ratios.
    a. Find each missing value. Explain your reasoning.
    b. Graph each set of points. Plot the points on a coordinate plane.

25.

x	1	3	5	7
y	4	12	20	?

26.

x	2	4	?	8
y	5	10	15	20

27.

x	2	4	8	16
y	3	6	?	24

28.

x	?	6	9	18
y	4	8	12	24

A multiplication table can also show ratios.

	1	2	3	4	5	6	7	8	9	10
1	1	2	③	4	⑤	6	7	8	9	10
2	2	4	6	8	10	12	14	16	18	20
3	3	6	9	12	15	18	21	24	27	30
4	4	8	12	16	20	24	28	32	36	40
5	5	10	15	20	25	30	35	40	45	50
6	6	12	18	24	30	36	42	48	54	60
7	7	14	21	28	35	42	49	56	63	70
8	8	16	24	32	40	48	56	64	72	80
9	9	18	27	36	45	54	63	72	81	90
10	10	20	30	40	50	60	70	80	90	100

29. Locate the 3 and the 5 in the first row (circled).

$$1 \times 3 = 3 \text{ and } 1 \times 5 = 5$$

An equivalent ratio to 3 : 5 can be found by looking above or below the two values in the chart. For example, 6 : 10 is an equivalent ratio.

   a. Use the chart to find two different equivalent ratios to 3 : 5.

   b. Kyle says 12 : 16 is equivalent to 3 : 5. Is he correct? Explain your reasoning.

30. Locate the numbers 15 and 30 in the fifth row.
   a. Use the chart to find two different equivalent ratios to 15 : 30.
   b. Tim says 27 : 54 is equivalent. Is he correct? Explain your reasoning.

31. Compare the ratios 1 : 3 and 2 : 5 by copying and completing the table at right. The number at the top of each column is the first value in the ratio and the number in each white cell is the second value in the ratio. If the first number in each ratio is 8, which ratio gives a larger second value? Explain your answer.

first value

	2	4	6	8
1:3	6	12		24
2:5	5	10	15	

second value

Tic-Tac-Toe ~ Body Ratios

There are some interesting ratios related to bodies. Choose whether to measure the lengths below using customary units or metric units. If you choose customary units, measure each length to the nearest inch. If you choose metric units, measure each length to the nearest centimeter.
- Measure and record your height. Measure from the floor to the top of your head (without shoes).
- Measure and record your wingspan. Extend your arms at a 90° angle from your body. Measure the length across your back from one fingertip to the other fingertip.
- Measure and record the length of your foot from the back of your heel to the tip of your longest toe.
- Measure and record the length of your forearm from the inner crease of your elbow along the inside of your arm to your wrist.

Step 1: Find the ratio of your height to your wingspan.
Step 2: Find the ratio of your foot to your forearm.
Step 3: Find the ratio of height to wingspan and the ratio of foot to forearm for four additional people. Record the measurements and ratios of your own measurements and four other people in a chart.
Step 4: Look at each group of ratios. Are the ratios similar to each other for $\frac{\text{height}}{\text{wingspan}}$? Are the ratios similar to each other for $\frac{\text{foot}}{\text{forearm}}$? Explain whether the ratios are closest to 1 : 1, 1 : 2, 2 : 1 or something else.
Step 5: Find two different lengths on your body to measure that would give you a ratio close to 1 : 2. Show the measurements and corresponding ratios in your explanation.

Lesson 1.1 ~ Ratios

GEOMETRIC SEQUENCES

LESSON 1.2

Recognize and complete geometric sequences.

Zane put a bacteria cell in a petri dish. The number of bacteria cells in the petri dish doubled every day.

Number of Days	Number of Bacteria Cells
1	1
2	$1 \times 2 = 2$
3	$2 \times 2 = 4$
4	$4 \times 2 = 8$
5	$8 \times 2 = 16$
6	$16 \times 2 = 32$
7	$32 \times 2 = 64$

After one week, there were 64 bacteria cells in Zane's petri dish. The number of bacteria cells can be written as a list of numbers: 1, 2, 4, 8, 16, 32, 64, … If the bacteria cells are left in the petri dish, they will continue to double every day.

An ordered list of numbers is called a **sequence** of numbers. Each number in the list is called a **term**. The petri dish example is a **geometric sequence** because each term is found by multiplying the previous term by the same number. The number you multiply by each time to get the next term is the ratio between the terms.

$$1 \xrightarrow{\times 2} 2 \xrightarrow{\times 2} 4 \xrightarrow{\times 2} 8 \xrightarrow{\times 2} 16$$

1st term   2nd term   3rd term   4th term   5th term

The ratio in Zane's petri dish between the terms was 2 or $\frac{2}{1}$. This is because each term was doubled or multiplied by 2. The ratio can also be determined by writing the ratio of each term to its previous term.

$$\frac{2^{nd} \text{ term}}{1^{st} \text{ term}} = \frac{2}{1} \qquad \frac{3^{rd} \text{ term}}{2^{nd} \text{ term}} = \frac{4}{2} = \frac{2}{1} \qquad \frac{4^{th} \text{ term}}{3^{rd} \text{ term}} = \frac{8}{4} = \frac{2}{1}$$

Since each ratio of one term to its previous term simplifies to $\frac{2}{1}$ or 2, the ratio of the geometric sequence is 2.

EXAMPLE 1

Find the ratio of each geometric sequence. Use the ratio to find the next two terms of the geometric sequence.
a. 1, 3, 9, 27, … b. 256, 64, 16, 4, …

SOLUTIONS

a. To find the ratio, divide one term by the previous term.

$\frac{3}{1} = 3$ $\frac{9}{3} = 3$ $\frac{27}{9} = 3$

The ratio is 3.

Find the next two terms by multiplying the previous term by 3.

$27 \times 3 = 81$ $81 \times 3 = 243$

The next two terms are 81 and 243.

b. To find the ratio, divide one term by the previous term.

$\frac{64}{256} = \frac{1}{4}$ $\frac{16}{64} = \frac{1}{4}$ $\frac{4}{16} = \frac{1}{4}$

The ratio is $\frac{1}{4}$.

Find the next two terms by multiplying the previous term by $\frac{1}{4}$.

$4 \times \frac{1}{4} = 1$ $1 \times \frac{1}{4} = \frac{1}{4}$

The next two terms are 1 and $\frac{1}{4}$.

EXAMPLE 2

One geometric sequence begins with the number 2. Another begins with the number 3. Each geometric sequence has a ratio of 5. Write the first five terms of each geometric sequence.

SOLUTION

Since the ratio of both geometric sequences is 5, multiply each term in the sequence by 5 to get the next term.

The geometric sequences are: 2, 10, 50, 250, 1250, … and 3, 15, 75, 375, 1875, …

EXPLORE! **NUMBER PATTERNS**

A. 2, 12, 72, 432, …   B. 3, 6, 9, 12, …   C. 64, 32, 16, 8, …

D. $\frac{1}{2}$, 1, $1\frac{1}{2}$, 2, …   E. $\frac{1}{2}$, $\frac{1}{4}$, $\frac{1}{8}$, $\frac{1}{16}$, …

Step 1: Find the next two numbers in each sequence (**A–E**) above.

Step 2: Explain how you found the next two numbers in each sequence.

EXPLORE! (CONTINUED)

Step 3: **a.** Identify which sequences (**A–E**) are geometric sequences.
b. Find the first term of each geometric sequence.
c. Find the ratio of each geometric sequence.

Step 4: An *arithmetic sequence* is formed when the same number is added to a term to find the next term.
a. Which sequences (**A–E**) are arithmetic sequences?
b. Do arithmetic sequences have the same ratio between terms? Explain using examples.

Step 5: Write the first five terms of your own geometric sequence. Identify the first term and the ratio of the sequence.

EXERCISES

Find the ratio of each geometric sequence. Use the ratio to find the next two terms.

1. 1, 5, 25, 125, …

2. 1, 10, 100, 1000, …

3. 800, 400, 200, 100, …

4. 3, 1, $\frac{1}{3}$, $\frac{1}{9}$, …

5. 3, $\frac{3}{2}$, $\frac{3}{4}$, $\frac{3}{8}$, …

6. 4, 12, 36, 108, …

7. A flu epidemic swept through the sixth grade. One day, when the epidemic was at its worst, 128 students were absent. Each day after that, half of the students who had been absent were well enough to return to school.
a. Write a geometric sequence showing the number of students absent each day after the worst of the epidemic. Use a first term of 128.
b. What is the ratio of the geometric sequence?
c. Will there ever be 0 students absent in this sequence? Explain your reasoning.

Write the first five terms of a geometric sequence given the first term and the ratio.

8. first term: 1
ratio: 3

9. first term: 10,000
ratio: $\frac{1}{10}$

10. first term: 1
ratio: 4

Determine whether or not each given sequence is a geometric sequence. If the sequence is a geometric sequence, find its ratio.

11. 1, 4, 8, 12, …

12. 3, 6, 12, 24, …

13. 2000, 200, 20, 2, …

14. 162, 54, 18, 6, …

15. 25, 5, 1, $\frac{1}{5}$, …

16. $\frac{1}{8}$, $\frac{2}{8}$, $\frac{3}{8}$, $\frac{4}{8}$, …

Lesson 1.2 ~ Geometric Sequences

17. Find a piece of rectangular paper you can fold. You will be folding this paper to create a sequence.
   a. The original sheet of paper is 1 rectangle. Write a 1 as the first number in your sequence.
   b. Fold the sheet of paper in half. Open it and record the number of non-overlapping rectangles created with the fold.
   c. Refold the paper in half. Fold the paper in half again. Open the paper and record the number of non-overlapping rectangles created with the folds.
   d. Repeat this process until you can no longer fold the paper in half. How many times could you fold the paper before you could no longer fold it in half?
   e. Is the sequence of numbers you created a geometric sequence? Explain your reasoning using ratios.

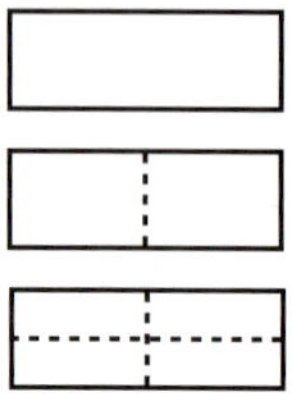

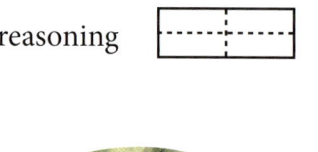

18. Svetlana told Marcos he had a choice about how much money he could earn for 30 days of work.
   Option #1: Marcos could earn $100 per day.
   Option #2: Marcos gets paid whatever his salary is at the end of 30 days. On the first day his salary is $0.01. Each day he comes to work his salary doubles.
   a. How much money would Marcos earn in 30 days with **Option #1**?
   b. Write the first five terms of the geometric sequence for **Option #2**.
   c. Use your calculator to find how much money Marcos would earn with **Option #2**. Find the value of the 30th term in the geometric sequence.
   d. Which option should Marcos choose? Explain your reasoning.

19. Write a sequence that is a geometric sequence. What is the ratio of the sequence?

20. Write a sequence that is NOT a geometric sequence. Explain why it is NOT a geometric sequence.

REVIEW

Write two different equivalent ratios for each ratio.

21. 6 : 9

22. 10 to 60

23. $\frac{24}{36}$

Write a ratio for each situation.

24. Eight out of 10 dog owners use leashes when walking their dogs.
   a. Write the ratio of dogs walking on leashes to total dogs walking.
   b. Write the ratio of dogs walking on leashes to dogs not walking on leashes.

25. Five red tulips and 7 yellow tulips are in a vase.
   a. Write the ratio of red tulips to yellow tulips in the vase.
   b. Write the ratio of red tulips to total tulips in the vase.
   c. Write the ratio of yellow tulips to total tulips in the vase.

26. Four out of six bikes parked in front of a school are black. Later in the day there are 33 bikes in front of the school with the same ratio of black bikes to total bikes. How many black bikes are now in front of the school? Show all work necessary to justify your answer.

Tic-Tac-Toe ~ Number Sequences

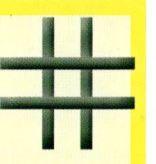

A **geometric sequence** is created by multiplying each term in the list of numbers by a ratio to find the next term.

Example: To find each term in the geometric sequence below multiply by a ratio of $\frac{3}{1} = 3$.

$$1, 3, 9, 27, 81, \ldots$$

There are other types of sequences that have a list of numbers with a pattern. One is an **arithmetic sequence**. To create an arithmetic sequence, you need to add the same number to each term to find the next term. This means there is the same difference between terms.

Example: To find each term in the arithmetic sequence below add 4 to each term.

$$3, 7, 11, 15, 19, 23, 27, 31, \ldots$$

Another type of sequence is called a **Fibonacci sequence**, named after one of the leading mathematicians in the Middle Ages. The Fibonacci sequence has a first and second term of 1. Each term after that is created by adding the previous two terms.

$$1, 1, 2, 3, 5, 8, 13, 21, \ldots$$

first term → 1
second term → 1
third term → 1 + 1 = 2
fourth term → 1 + 2 = 3

fifth term → 2 + 3 = 5
sixth term → 3 + 5 = 8
seventh term → 5 + 8 = 13
eighth term → 8 + 13 = 21

Step 1: Create two different geometric sequences, one with a ratio that is a whole number and one with a ratio that is a fraction less than 1. Explain why each sequence is a geometric sequence.

Step 2: Create two different arithmetic sequences, one with a difference between terms that is a whole number and one with a difference between terms that is a fraction or a mixed number. Explain why each sequence is an arithmetic sequence.

Step 3: Research the Fibonacci sequence and explain the rabbit problem used to create this list of numbers. Include at least one place in nature where the numbers from a Fibonacci sequence appear.

Step 4: Create your own sequence and explain the pattern used to find each term.

Lesson 1.2 ~ Geometric Sequences

CUSTOMARY CONVERSIONS

LESSON 1.3

🎯 Convert customary measurements.

The highest altitude in America is Mt. McKinley in Alaska at 20,320 feet. The lowest is Death Valley in California at 282 feet below sea level.

In 2009, the tallest building in the United States was the Willis Tower in Chicago, Illinois at 1,450 feet.

The deepest lake in the United States is Crater Lake in Southern Oregon at 1,996 feet.

These are examples of objects measured using feet, one of the customary units. Inches, feet, yards and miles are customary units for measuring length. Customary units also include ounces, tons and pounds for weight and cups, pints, quarts and gallons for volume. The customary system of measurement is used in the United States.

Measurements can be renamed. For example, once an object measures 12 inches, it can be renamed as 1 foot. The process of renaming a measurement using different units is called **conversion**.

Some common conversions are shown below for customary units.

Length	Weight	Volume	Time
1 foot (*ft*) = 12 inches (*in*) 1 yard (*yd*) = 3 feet (*ft*) 1 mile (*mi*) = 5,280 feet (*ft*)	1 pound (*lb*) = 16 ounces (*oz*) 1 ton = 2,000 pounds (*lbs*)	1 cup = 8 fluid ounces (*oz*) 1 pint (*pt*) = 2 cups 1 quart (*qt*) = 2 pints (*pt*) 1 gallon (*gal*) = 4 quarts (*qt*)	1 minute (*min*) = 60 seconds (*sec*) 1 hour (*hr*) = 60 minutes (*min*) 1 day = 24 hours (*hrs*) 1 week = 7 days 1 year = 52 weeks 1 year = 12 months

It is helpful when estimating to have a general idea how long, large or heavy these different measurements are.

EXAMPLE 1

Choose a reasonable estimate. Explain your reasoning for each choice.
a. amount of water in a personal water bottle: 1 quart or 1 gallon
b. length of a couch: 6 yards or 6 feet
c. weight of an elephant: 2 pounds or 2 tons

SOLUTIONS

a. <u>1 quart.</u> Think about a gallon of milk. That is larger than a personal water bottle.

b. <u>6 feet.</u> People are measured in feet. A 6 foot person could lay on a couch. Think about yards on a football field. Six yards would make a very long couch.

c. <u>2 tons.</u> Elephants are quite large. Most human babies weigh more than 2 pounds when they are born.

14   Lesson 1.3 ~ Customary Conversions

Sometimes it is necessary to convert from one unit to another. Smaller units can be converted to larger units, like inches to feet. Larger units can be converted to smaller units, like days to hours.

EXAMPLE 2

Using the information in the conversion table, find the number of yards in 36 feet.

Feet	3	6	9	12	24	36
Yards	1	2	3	4	8	?

SOLUTION

Each pair of feet and yard values shows the ratio of feet to yards. Choose any ratio and write it as a fraction. Simplify the ratio, if needed.

$$\frac{3 \text{ feet}}{1 \text{ yard}}$$

Find an equivalent ratio with 36 feet.

$$\frac{3 \text{ feet}}{1 \text{ yard}} \xrightarrow{\times 12} \frac{36 \text{ feet}}{12 \text{ yards}}$$

There are 12 yards in 36 feet.

UNIT CONVERSIONS

When converting to *smaller* units, **MULTIPLY** (×).
When converting to *larger* units, **DIVIDE** (÷).

EXAMPLE 3

Complete each conversion.
a. 2 miles = _____ feet
b. 280 days = _____ weeks
c. 18 inches = _____ feet

SOLUTIONS

a. Find the conversion factor.
Multiply 2 miles by 5,280.
2 miles = 10,560 feet.

1 mile = 5,280 feet
2 × 5280 = 10560

When converting to a smaller unit, multiply.

b. Find the conversion factor.
Divide 280 by 7.
280 days = 40 weeks.

1 week = 7 days
280 ÷ 7 = 40

When converting to a larger unit, divide.

c. Find the conversion factor.
Divide 18 by 12.
18 inches = 1.5 feet.

1 foot = 12 inches
18 ÷ 12 = 1.5

EXERCISES

Choose a reasonable estimate. Explain your reasoning for each choice.

1. length of a cell phone: 4 inches or 4 feet

2. weight of a bag of flour: 5 ounces or 5 pounds

Lesson 1.3 ~ Customary Conversions

Choose a reasonable estimate. Explain your reasoning for each choice.

3. time to walk 1 mile: 15 minutes or 15 seconds

4. length of a football field: 100 yards or 100 feet

5. amount of soda in a glass: 8 pints or 8 ounces

Find the missing value in each conversion table. Explain your reasoning.

6.
quarts	4	8	16	?
gallons	1	2	4	6

7.
inches	12	24	36	48
feet	1	2	?	4

Plot the points in each conversion table. Use the graph to find the missing value.

8.
feet	3	6	12	15
yards	1	2	4	?

9.
cups	2	4	8	12
pints	1	2	4	?

Complete each conversion. Show all work necessary to justify your answer.

10. 6 feet = _____ inches

11. 2 gallons = _____ quarts

12. 48 inches = _____ feet

13. 5,000 pounds = _____ tons

14. 5 weeks = _____ days

15. 14 yards = _____ feet

16. 24 pints = _____ quarts

17. 30 inches = _____ feet

18. 4 tons = _____ pounds

19. 5 hours = _____ minutes

20. Crater Lake is 1,996 feet deep.
    a. How many *inches* deep is Crater Lake?
    b. How many *miles* deep is Crater Lake? Round your answer to the nearest tenth.

21. The Willis Tower in Chicago is 1,450 feet tall. What is the height in inches?

22. Mikayla and Shawna figured out how many seconds are in 2 hours.

Mikayla	Shawna
1 hour = 60 minutes 2 hours = 2 × 60 = 120 2 hours = 120 seconds	1 minute = 60 seconds 1 hour = 60 minutes 1 hour = 60 × 60 = 3,600 seconds 2 hours = 2 × 3,600 = 7,200 seconds 2 hours = 7,200 seconds

    a. Which person is correct?
    b. What unit should be at the end of the number 120 in Mikayla's work?
    c. Finish Mikayla's work. Check your work by looking at Shawna's answer.

Lesson 1.3 ~ Customary Conversions

REVIEW

Simplify each ratio. Write each ratio as a fraction, with a colon and using the word "to".

23. 8 : 16

24. 10 to 30

25. $\frac{9}{12}$

Find the ratio of each geometric sequence. Use the ratio to find the next two terms.

26. 5, 20, 80, 320, ...

27. 8, 4, 2, 1, ...

28. 5000, 1000, 200, 40, ...

29. $\frac{1}{6}$, 1, 6, 36, ...

Tic-Tac-Toe ~ Gnat Story

Two gnats were sitting on one side of a room 512 inches long. One gnat decided to fly to the other side of the room. Since he was tired from a long day, his friend suggested that he fly half the distance to the wall, then stop to rest. Then he could fly half the remaining distance and stop to rest again. The friend suggested he continue this pattern until he reached the wall on the other side of the room. The gnat agreed and began his journey.

Step 1: Find the length of the room the gnat needs to fly before he can rest for the first time.
Step 2: Find the length of the room the gnat needs to fly before he can rest for the second time.
Step 3: Find the length of the room the gnat needs to fly before he can rest for the third time.
Step 4: Continue recording the length of the room the gnat needs to fly for each of his next ten trips.
Step 5: List the lengths of the room the gnat needs to fly from #1-4 in a geometric sequence.
Step 6: Identify the first term and the ratio of the geometric sequence.
Step 7: Find the next five terms of the geometric sequence.
Step 8: Will the geometric sequence ever have a term equal to 0? Explain your reasoning.
Step 9: If the gnat continues to fly half the remaining distance to the wall before resting, will he ever reach the wall? Explain your reasoning in a full paragraph.

Tic-Tac-Toe ~ Height Conversions

Find the height of ten different landmarks in the United States not already given in this textbook.
Example: Find the height of Mt. Whitney in California, the height of Olo'upena Falls in Hawaii, the depth of the Grand Canyon or the depth of Lake Michigan.
Record each measurement in feet. Convert each measurement to inches, yards and miles. Round each answer to the nearest tenth, if necessary. Copy and complete the table below with each of your ten landmarks. The last column is for your own height.

Name of Landmark											You
Height (*ft*)											
Height (*in*)											
Height (*yd*)											
Height (*mi*)											

METRIC CONVERSIONS

LESSON 1.4

Convert metric measurements.

The metric system of measurement is used all over the world. The metric system is a base-ten, or decimal, system of measurement.

All metric units of length are defined in terms of the meter. Metric units of mass are based on the gram. Metric units of volume are defined in terms of the liter. The prefix on a measurement relates the unit to a meter, gram or liter.

Prefixes for Metric Measurements

Prefix	Meaning using Words	Meaning using Numbers
milli	one-thousandth	$\frac{1}{1000}$ or 0.001
centi	one-hundredth	$\frac{1}{100}$ or 0.01
deci	one-tenth	$\frac{1}{10}$ or 0.1
deca	ten	10
hecto	one hundred	100
kilo	one thousand	1000

A <u>centi</u>meter is one-hundredth of a meter. A <u>kilo</u>meter is 1,000 times larger than a meter. Also, a <u>milli</u>liter is one-thousandth of a liter and a <u>deca</u>gram is 10 times heavier than a gram.

Certain situations require converting from one metric unit to another. Some common conversions for metric units are shown below.

Length	Mass	Volume
1 centimeter (*cm*) = 10 millimeters (*mm*) 1 meter (*m*) = 100 centimeters (*cm*) 1 kilometer (*km*) = 1,000 meters (*m*)	1 gram (*g*) = 1,000 milligrams (*mg*) 1 kilogram (*kg*) = 1,000 grams (*g*)	1 liter (*L*) = 1,000 milliliters (*mL*)

When estimating, it is helpful to have a general idea of the size of the common metric units.

18   Lesson 1.4 ~ Metric Conversions

EXAMPLE 1 Choose a reasonable estimate. Explain your reasoning for each choice.
a. weight of a paperclip: 1 gram or 1 kilogram
b. length of a van: 4 kilometers or 4 meters
c. water in a personal water bottle: 1 milliliter or 1 liter

SOLUTIONS

a. <u>1 gram</u>. It is helpful to estimate the weight of 1 gram by thinking of 1 gram as the weight of a paperclip.

b. <u>4 meters</u>. A meter is a little longer than a yard which is three feet. So 4 meters is slightly longer than 12 feet.

c. <u>1 liter</u>. One liter could be the size of a large personal water bottle. A drop of water would be a good estimate for 1 milliliter.

EXAMPLE 2 Using the information in the conversion table, find the number of centimeters in 60 millimeters.

millimeters	10	20	30	50	60
centimeters	1	2	3	5	?

SOLUTION

Each pair of millimeter and centimeter values show the ratio of millimeters to centimeters. Choose any ratio and write it as a fraction. Simplify the ratio, if needed.

$$\frac{10\ mm}{1\ cm}$$

Find an equivalent ratio with 60 millimeters.

$$\frac{10\ mm}{1\ cm} \xrightarrow{\times 6} \frac{60\ mm}{6\ cm}$$

There are 6 centimeters in 60 millimeters.

When converting from one metric unit to another, you will multiply or divide by 10, 100, 1000, ...

EXAMPLE 3 Complete each conversion.
a. 2 kilometers = _____ meters
b. 3.9 kilograms = _____ grams
c. 4,000 meters = _____ kilometers
d. 650 centimeters = _____ meters

SOLUTIONS

a. Find the conversion factor.   1 kilometer = 1000 meters
   Multiply 2 kilometers by 1,000.   2 × 1000 = 2000
   2 kilometers = 2,000 meters.

b. Find the conversion factor.   1 kilogram = 1000 grams
   Multiply 3.9 kilograms by 1,000.   3.9 × 1000 = 3900
   3.9 kilograms = 3,900 grams.

c. Find the conversion factor.   1 kilometer = 1000 meters
   Divide 4,000 meters by 1,000.   4000 ÷ 1000 = 4
   4,000 meters = 4 kilometers.

d. Find the conversion factor.   1 meter = 100 centimeters
   Divide 650 centimeters by 100.   650 ÷ 100 = 6.5
   650 centimeters = 6.5 meters.

Lesson 1.4 ~ Metric Conversions

EXERCISES

Choose a reasonable estimate. Explain your reasoning for each choice.

1. length of a CD music case: 13 millimeters or 13 centimeters

2. weight of a pen: 3 grams or 3 kilograms

3. length of a baseball bat: 1 meter or 1 decimeter

4. length of a laptop computer: 38 centimeters or 38 meters

5. Liv says if two positive measurements in meters and kilometers are equivalent, the number in front of meters will be larger than the number in front of kilometers. Is she correct? Explain your reasoning.

6. Which of the lengths below equal 50 meters? Write each answer that applies and explain how you know your answer is correct.

    5,000 centimeters   50,000 millimeters   0.5 kilometers   500 decimeters

Find the missing value in each conversion table. Explain your reasoning.

7.
milligrams	1,000	2,000	4,000	7,000
grams	1	2	4	?

8.
centimeters	200	400	600	?
meters	2	4	6	45

Plot the points in each conversion table. Use the graph to find the missing value.

9.
millimeters	10	20	40	?
centimeters	1	2	4	5

10.
meters	1,000	2,000	5,000	?
kilometers	1	2	5	8

Complete each conversion. Show all work necessary to justify your answer.

11. 6 meters = _____ centimeters

12. 2 liters = _____ milliliters

13. 5,000 meters = _____ kilometers

14. 10 kilograms = _____ grams

15. 3 liters = _____ milliliters

16. 15 millimeters = _____ centimeters

17. 7 kilometers = _____ meters

18. 450 milliliters = _____ liters

19. 92,000 grams = _____ kilograms

20. 2 kilometers = _____ centimeters

21. A gym is 18 meters tall. How tall is the gym in centimeters?

22. Rewrite 4 meters using two different metric units. Use mathematics to justify your answer.

23. If you could only measure objects using customary units or metric units, which would you choose? Explain your reasoning.

REVIEW

Complete the conversion. Show all work necessary to justify your answer.

24. 5 feet = _____ inches

25. 3 tons = _____ pounds

26. 4 pints = _____ cups

Find the ratio for each geometric sequence. Use the ratio to find the next term in each geometric sequence.

27. 10, 100, 1000, 10000, …

28. $\frac{3}{256}, \frac{3}{64}, \frac{3}{16}, \frac{3}{4}, \ldots$

Tic-Tac-Toe ~ Metric Lengths

Signs along freeways in the United States tell you the number of miles until the next exit or city. Mountains and waterfalls are measured in feet. People talk about how many inches children have grown. These are customary measurements. People in most countries around the world see signs indicating kilometers, measure mountains and waterfalls using meters and discuss the growth of children in terms of centimeters.

Step 1: Research to find the countries that use customary measurement and the countries that use metric measurement. List 3 countries that use customary measurement. List ten countries that use metric measurement.

Step 2: Find the height, in meters, of one landmark in five different countries that use the metric system of measurement found in **Step 1**. Record each measurement in meters. Convert each measurement to millimeters, centimeters and kilometers. Round all measurements to the nearest tenth. Copy and complete the table below with each of your five landmarks. Include the name of the landmark as well as the country where it is located.

Name of Landmark					
Height (*m*)					
Height (*mm*)					
Height (*cm*)					
Height (*km*)					

Tic-Tac-Toe ~ Student Survey

Determine a question you would like to ask 40 people. You could ask people their favorite movie type, music type, food type or any other preference of your choosing. Write this question on your paper.

You will need three, four or five different possible answers to the question you chose.

Example: Ask students, "Which type of ice cream is your favorite ~ chocolate, vanilla, strawberry, cookies and cream or fudge swirl?"

Record the responses to your question in a chart like the one below.

Types of _____					
Frequency (number of students who chose each type)					

Step 1: For each preference find:
 ♦ the ratio of students who preferred that choice to those who did not prefer that choice.
 ♦ the ratio of students who preferred that choice to all students surveyed.

Step 2: Clearly organize and display your question as well as your table with responses and ratios.

PERIMETER AND AREA

LESSON 1.5

🎯 Convert measurements to find perimeter and area.

The **perimeter** of a shape is the total length around a shape. Add the lengths of the sides of the shape together to find the perimeter. The **area** of a shape is the number of square units that fit inside the shape.

EXPLORE!   THE PATIO

Trent and Mandy put a rectangular patio in their back yard. They plan to put a fence around the patio. They will use square tiles that are 1 meter by 1 meter to fill in the patio.

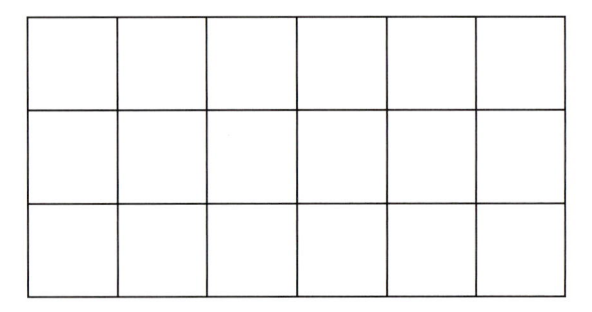

Step 1: Find the perimeter of the patio so Trent and Mandy know how much fencing to buy. Your answer should be in meters.

Step 2: Find the area of the patio. Your answer should be in square meters.

Step 3: Convert the measurements of each tile to centimeters.
   a. Find the perimeter of the patio in centimeters.
   b. Find the area of the patio in square centimeters.

Step 4: Find the ratio of the perimeter of the patio in meters to the perimeter of the patio in centimeters.

Step 5: Find the ratio of the area of the patio in meters to the area of the patio in centimeters.

Step 6: Are the ratios from **Steps 4 and 5** equal? Why or why not?

There are times when it will be necessary to convert from one unit of measurement to another when finding area or perimeter.

feet ↔ inches   *yards ↔ feet*

centimeters ↔ meters

22 Lesson 1.5 ~ Perimeter and Area

EXAMPLE 1

Find the perimeter of the rectangle in:
a. meters.
b. centimeters.

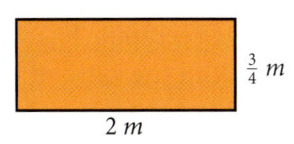

SOLUTIONS

a. Add the length of all four sides.
The perimeter is $5\frac{1}{2}$ meters.

$2 + \frac{3}{4} + 2 + \frac{3}{4} = 5\frac{1}{2}$

> Remember to write the name of the units in your answer.

b. Convert meters to centimeters using 1 meter = 100 centimeters.

$2 \times 100 = 200 \, cm$
$\frac{3}{4} \times 100 = 75 \, cm$

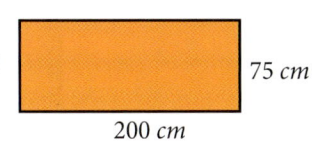

Add the new lengths of all four sides.
The perimeter is 550 centimeters.

$200 + 75 + 200 + 75 = 550$

EXAMPLE 2

Find the area of the triangle in:
a. square feet.
b. square inches.

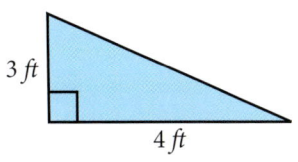

SOLUTIONS

a. Use the area formula for a triangle.

$A = \frac{1}{2} \times b \times h$

Locate the length of the base and height.

$b = 4$
$h = 3$

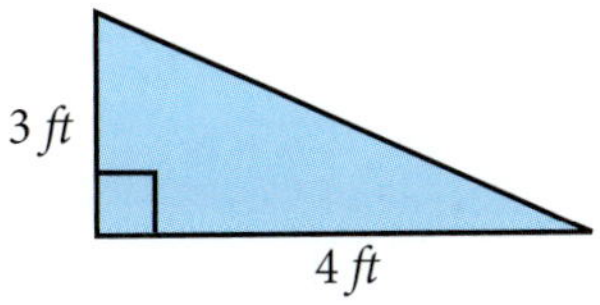

Calculate the area.

$A = \frac{1}{2} \times 4 \times 3 = 6$

The area is 6 square feet.

b. Convert feet to inches using 1 foot = 12 inches.

$b = 4 \times 12 = 48 \, in$
$h = 3 \times 12 = 36 \, in$

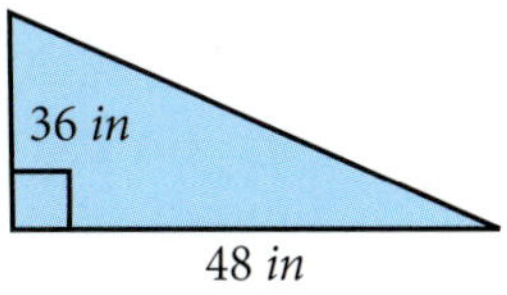

Calculate the area using the new lengths.

$A = \frac{1}{2} \times 48 \times 36 = 864$

The area is 864 square inches.

Area and perimeter are used in many careers. Interior designers must find area to determine how much wall paper is needed to cover a wall. Construction workers use perimeter to determine the length of baseboards needed to go around an entire room. People may need to convert between units if the materials they purchase are not in the same unit they used for measuring.

Lesson 1.5 ~ Perimeter and Area

EXAMPLE 3

One rectangle has sides twice as long as another rectangle, as shown below.
a. Find the ratio of the smaller perimeter to the larger perimeter.
b. Find the ratio of the smaller area to the larger area.

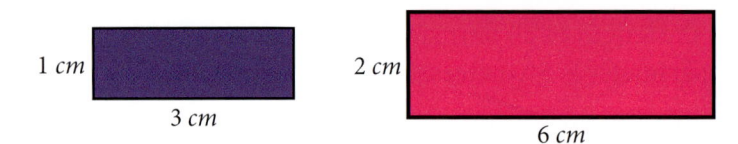

1 cm   2 cm
3 cm   6 cm

SOLUTIONS

a. Find the perimeter of the smaller rectangle. $1 + 3 + 1 + 3 = 8$ cm

Find the perimeter of the larger rectangle. $2 + 6 + 2 + 6 = 16$ cm

Write the ratio of the smaller perimeter $\frac{8}{16} = \frac{1}{2}$ or $1:2$
to the larger perimeter.

b. Find the area of the smaller rectangle. $3 \times 1 = 3$ square centimeters

Find the area of the larger rectangle. $6 \times 2 = 12$ square centimeters

Write the ratio of the smaller area $\frac{3}{12} = \frac{1}{4}$ or $1:4$
to the larger area.

EXERCISES

1.
8 ft
8 ft

a. Find the perimeter of the square in feet.
b. Find the perimeter of the square in inches.

2.
20 cm   29 cm
21 cm

a. Find the perimeter of the triangle in centimeters.
b. Find the perimeter of the triangle in millimeters.

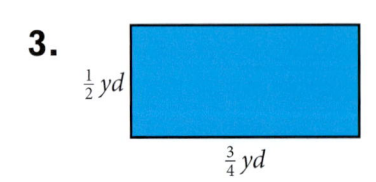

3.
$\frac{1}{2}$ yd
$\frac{3}{4}$ yd

a. Find the perimeter of the rectangle in yards.
b. Find the perimeter of the rectangle in feet.

4.
5 m   5 m
4 m
6 m

a. Find the perimeter of the triangle in meters.
b. Find the perimeter of the triangle in centimeters.

Lesson 1.5 ~ Perimeter and Area

5.

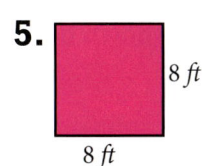

8 ft / 8 ft

a. Find the area of the square in square feet.
b. Find the area of the square in square inches.

6.

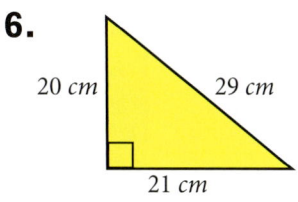

20 cm, 29 cm, 21 cm

a. Find the area of the triangle in square centimeters.
b. Find the area of the triangle in square millimeters.

7.

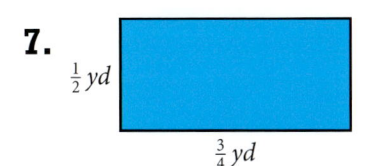

$\frac{1}{2}$ yd / $\frac{3}{4}$ yd

a. Find the area of the rectangle in square yards.
b. Find the area of the rectangle in square feet.

8.

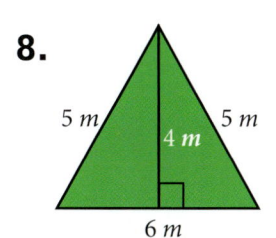

5 m, 5 m, 4 m, 6 m

a. Find the area of the triangle in square meters.
b. Find the area of the triangle in square centimeters.

9. How many square centimeters are in one square meter? Explain how you know your answer is correct.

10. Carlos says there are 3 square feet in one square yard. Is he correct? Explain your reasoning.

11. Kyler carpeted a bedroom in his house. The rectangular room is 12 ft by 15 ft.
   a. Find the area of the bedroom in square feet.
   b. The carpet is sold by the square yard. How many square yards is Kyler's bedroom?
   c. The carpet costs $41.95 per square yard. How much did the carpet cost?

12. Measure the sides of the rectangle to the nearest half centimeter using a metric ruler.

   a. Use your measurements to find the perimeter.
   b. Use your measurements to find the area.

13. Paul and Debbie put a fence around their swimming pool. The area they enclosed was a rectangle that measures 9 yards by 20 yards.
   a. The fencing is sold by the foot. What is the perimeter of the rectangle in feet? Use mathematics to justify your answer.
   b. The fencing costs $12 per foot. How much did the fencing cost?

14. Start with a square that is 1 inch on each side.
   a. Find the perimeter of the square.
   b. Double the sides of the square. Find the new perimeter.
   c. Double the sides of the square in **part b**. Find the new perimeter.
   d. Double the sides of the square in **part c**. Find the new perimeter.
   e. Double the sides of the square in **part d**. Find the new perimeter.
   f. List your answers from **parts a–e** as a sequence of numbers. Do the numbers create a geometric sequence? Use mathematics to justify your answer.

Lesson 1.5 ~ Perimeter and Area

15. One triangle has side lengths that are $1\frac{1}{2}$ times as large as the other triangle's side lengths, as shown below.

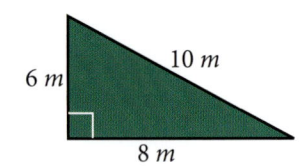

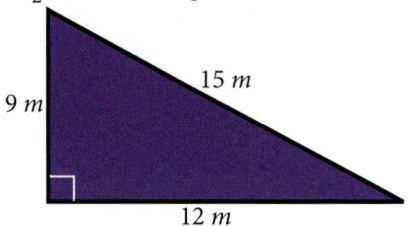

    a. Find the perimeter of each triangle.
    b. Find the ratio of the smaller perimeter to the larger perimeter.
    c. Find the area of each triangle.
    d. Find the ratio of the smaller area to the larger area.

16. Shelby and Ed built a deck in front of their house. It is a rectangular shape with sides measuring 10 ft by $10\frac{1}{2}$ ft.
    a. Find the perimeter of the deck.
    b. Find the area of the deck.

REVIEW

Find the ratio of each geometric sequence. Use the ratio to find the next term in the sequence.

17. 30, 15, 7.5, 3.75, …

18. 343, 49, 7, 1, $\frac{1}{7}$, …

19. 5, 10, 20, 40, …

20. Write the first five terms of a geometric sequence with first term 1 and ratio 1. What do you notice?

TIC-TAC-TOE ~ CARPET COST

You need to put new carpeting in your bedroom and two other rooms in your house. Complete each step to find the cost for the carpeting.

Step 1: Measure your bedroom and two other rooms in your house to the nearest foot. Draw a sketch of the shape of the floor for each room and label it with your measurements.

For example: (rectangle 12 ft by 10 ft)

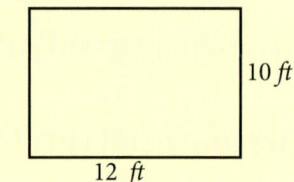

Step 2: You like two different carpets. You want to purchase the least expensive one. The brown carpet costs $2.75 per square foot. The white carpet costs $22.00 per square yard. Explain how you can determine which carpet to buy.

Step 3: Find the total cost to carpet the three rooms in your house with the least expensive carpet.

TIC-TAC-TOE ~ SQUARE UNITS

People often get confused when changing units on problems involving area. Design a poster that explains how to convert each of the following units using pictures and words. Include an example for each one.
- square yards to square feet
- square feet to square inches
- square meters to square centimeters
- square centimeters to square millimeters

TIC-TAC-TOE ~ PAY RAISE

You have just been hired for a new job as a shipping clerk. Your starting salary is $40,000 per year. Your boss has told you there are two different plans you can choose from to get your raise each year.

Plan A: Every year your pay will increase by $5,000 until your tenth year.
Plan B: Every year your pay will increase by $\frac{1}{10}$ of your previous year's salary until your tenth year.

1. Copy and complete the table to record the salaries you could earn with each plan for your first ten years with the company.

Year	1	2	3	4	5	6	7	8	9	10
Plan A	$40,000	$45,000								
Plan B	$40,000	$44,000								

Plan A
Year 1 → $40,000
Year 2 → $40,000 + $5,000 = $45,000

Plan B
Year 1 → $40,000
Year 2 → $40,000 × $\frac{1}{10}$ = $4,000 (this is the raise)
$40,000 + $4,000 = $44,000

2. If you intend to stay with the company five years, which plan should you choose? Explain your reasoning.

3. If you intend to stay with the company ten years, which plan should you choose? Explain your reasoning.

4. List the numbers as a sequence for Plan B.
$$40000, 44000, \ldots$$
These numbers make a geometric sequence. Find the ratio of the geometric sequence. Explain what this ratio means.

Lesson 1.5 ~ Perimeter and Area

REVIEW — BLOCK 1

Vocabulary

area
conversion
geometric sequence
perimeter
ratio
term
sequence

- Simplify and write ratios three ways.
- Recognize and complete geometric sequences.
- Convert customary measurements.
- Convert metric measurements.
- Convert measurements to find perimeter and area.

Lesson 1.1 ~ Ratios

Simplify each ratio. Write each ratio as a fraction, with a colon and using the word "to".

1. 8 to 16

2. $\frac{9}{24}$

3. 4 : 2

4. Write two different equivalent ratios to 5 : 6.

5. Compare the number of stars to the number of squares. If the ratio stays the same and there are 20 stars, how many squares will there be? ★ ★ ★ ★   ■ ■ ■

Each table is made of *x*- and *y*-coordinates that form equivalent ratios.
  a. Find each missing value. Explain your reasoning.
  b. Graph each set of points. Plot the points on a coordinate plane.

6.

x	1	4	8	12
y	3	12	24	?

7.

x	4	8	?	16
y	5	10	15	20

8. Compare the ratios 1 : 4 and 1 : 5 by copying and completing the table below. The number at the top of each column is the first value in the ratio and the number in each white cell is the second value in the ratio. If the first number in each ratio is 12, which ratio gives a larger second value? Explain how you know your answer is correct.

	3	6	9	12
1 : 4	12	24	36	
1 : 5	15	30		60

Write a ratio in simplest form for each situation.

9. At recess 14 girls and 18 boys played dodge ball.
   a. Write the ratio of girls playing dodge ball to boys playing dodge ball.
   b. Write the ratio of girls playing dodge ball to all students playing dodge ball.
   c. Write the ratio of boys playing dodge ball to all students playing dodge ball.

10. There are 8 yellow carnations and 4 red roses in a bouquet.
   a. Write the ratio of yellow carnations to red roses.
   b. Write the ratio of yellow carnations to all flowers in the bouquet.
   c. Write the ratio of red roses to all flowers in the bouquet.

11. Fifteen students were asked which flavor of ice cream they liked best. Eight chose chocolate, 6 chose strawberry and 1 chose vanilla.
   a. Write the ratio of students who liked chocolate ice cream the best to total students asked.
   b. Write the ratio of students who liked strawberry ice cream best to those who did not choose strawberry ice cream.

12. There were 50,000 fans at a football game. Of those, 15,000 fans were wearing purple and the rest were wearing silver. Write three ratios using numbers and words with this information.

Lesson 1.2 ~ Geometric Sequences

13. Each time a person smiles, three more people smile. Suppose one person smiles.
   a. How many additional people will smile after the first person?
   b. How many additional people will smile after the people in **part a** smile?
   c. How many additional people will smile after the people in **part b** smile?
   d. Write the numbers generated in **parts a-c** in a sequence followed by the original number for the first person who smiled. If this sequence continues, will it be geometric? Use mathematics to justify your answer.

Find the ratio of each geometric sequence. Use the ratio to find the next two terms of each geometric sequence.

14. 10000, 1000, 100, 10, …

15. $1, \frac{1}{3}, \frac{1}{9}, \frac{1}{27}, \ldots$

16. 1, 20, 400, 8000, …

Write the first five terms of a geometric sequence given the first term and the ratio.

17. first term: 3
   ratio: 2

18. first term: 200
   ratio: $\frac{1}{2}$

19. first term: 10
   ratio: 3

Determine whether or not each sequence is a geometric sequence. If the sequence is a geometric sequence, find its ratio.

20. 4, 12, 36, 108, …

21. 5, 10, 15, 20, …

22. $4, 1, \frac{1}{4}, \frac{1}{16}, \ldots$

Lesson 1.3 ~ Customary Conversions

Choose a reasonable estimate. Explain your reasoning for each choice.

23. the height of Yosemite Falls: 2,425 miles or 2,425 feet

24. the weight of a car: 1 pound or 1 ton

25. the amount of water used to water a house plant: 2 cups or 2 quarts

Find the missing value in each conversion table.

26.

minutes	60	120	240	360
hours	1	2	?	6

27.

days	14	21	35	?
weeks	2	3	5	8

Complete each conversion. Show all work necessary to justify your answer.

28. 10 feet = _____ inches

29. 11 cups = _____ pints

30. 6 tons = _____ pounds

31. 2 miles = _____ feet

32. 120 hours = _____ days

33. 12 feet = _____ yards

34. Don knows there are 3 feet in 1 yard. He uses that to determine that there are 15 yards in 5 feet. Is Don correct? Explain your reasoning.

35. Which of the times below equal 1 year? Write each answer that equals 1 year.

    365 days   12 months   7 days   24 hours

Lesson 1.4 ~ Metric Conversions

Choose a reasonable estimate. Explain your reasoning for each choice.

36. weight of 5 paperclips: 5 kilograms or 5 grams

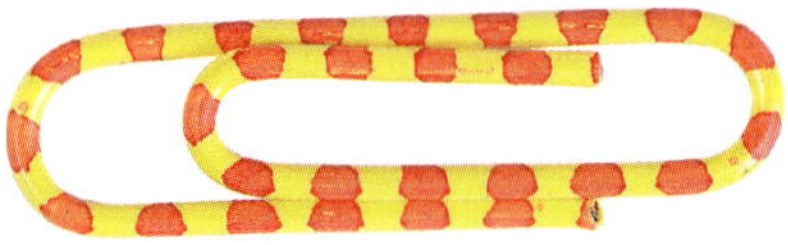

37. length of a piece of notebook paper: 28 centimeters or 28 millimeters

38. amount of water consumed by a football player during practice: 1 liter or 1 milliliter

39. Which of the lengths below equal 60 meters? Write each answer that applies and explain how you know your answers are correct.

    600 centimeters   60,000 millimeters   0.06 kilometers   600 decimeters

Find the missing value in each conversion table.

40.

millimeters	3,000	6,000	7,000	8,000
meters	3	6	?	8

41.

decimeters	10	30	50	?
meters	1	3	5	7

Block 1 ~ Review

Complete each conversion. Show all work necessary to justify your answer.

42. 17 centimeters = _____ millimeters

43. 800 meters = _____ kilometers

44. 80 centimeters = _____ meters

45. 2 kiloliters = _____ liters

46. 2000 grams = _____ kilograms

47. 1 kilometer = _____ centimeters

Lesson 1.5 ~ Perimeter and Area

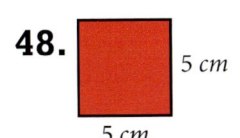

48. (square, 5 cm by 5 cm)
  a. Find the perimeter in centimeters.
  b. Find the perimeter in millimeters.

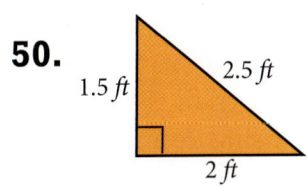

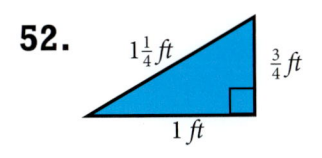

49. (rectangle, $\frac{2}{3}$ yd by $\frac{1}{2}$ yd)
  a. Find the perimeter in yards.
  b. Find the perimeter in feet.

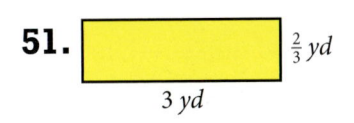

50. (right triangle, 1.5 ft, 2.5 ft, 2 ft)
  a. Find the perimeter in feet.
  b. Find the perimeter in inches.

51. (rectangle, 3 yd by $\frac{2}{3}$ yd)
  a. Find the area in square yards.
  b. Find the area in square feet.

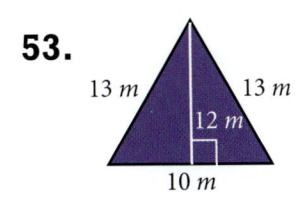

52. (right triangle, $1\frac{1}{4}$ ft, $\frac{3}{4}$ ft, 1 ft)
  a. Find the area in square feet.
  b. Find the area in square inches.

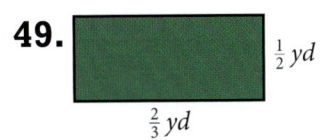

53. (triangle, 13 m, 13 m, base 10 m, height 12 m)
  a. Find the area in square meters.
  b. Find the area in square centimeters.

54. Edgar measured the dimensions of a square tile in millimeters. The tile was 120 millimeters on each side.
  a. Find the perimeter of the tile in millimeters and the area of the tile in square millimeters.
  b. The tile store sells the same tiles, but the dimensions are in centimeters. Find the perimeter and area of the tile in centimeters and square centimeters. Show all work necessary to justify your answers.

55. Carrie made a quilt that was 6 square yards. What are two different possible dimensions of the quilt using feet?

CAREER FOCUS

**ELLEN
REGISTERED NURSE**

I am a registered nurse. I run my own health consulting business. I help people get healthy, stay healthy and slow down the process of aging. Not very many registered nurses run their own business. Some nurses work in research and teach other nurses. Most nurses work in hospitals, clinics or doctor's offices. Some nurses specialize in certain types of illnesses such as heart disease or cancer. Other nurses care for a specific age, such as the elderly or children. In addition, some nurses choose to practice nursing through education programs designed to help people prevent disease or in nursing education programs. Since I do much of my health consulting by telephone and email, I work with people from all over the United States but am still able to live in historic Jacksonville, a small, charming town in southern Oregon.

As a registered nurse I use math in many ways. I use fractions and division to determine when clients are to take medication and supplements throughout the day. I use proportions to make sure that children take the right amount of milligrams of medication based on their current body weight (*mg/kg* of weight). I use addition and fractions to write up invoices for customers. I also use math on a daily basis to manage my business.

The best way to become a registered nurse is to first get a Bachelor of Science degree in Nursing (BSN). This can take 4 or 5 years depending on the requirements of individual programs. I received my BSN after 4 years of college. After that, I had to pass a test to become licensed to practice nursing. A nurse must have a current license in the state in which he or she is practicing.

Nurses just beginning their careers often work as interns for a period immediately after college graduation. Starting salaries can range from $30,000 to $50,000 per year. Nurses who run their own businesses or become managers or professors often earn $75,000 or more a year. Salary amounts vary depending on the type of responsibilities they have and whether they live in a large city or a rural area.

I became a nurse because I love people and wanted to help them stay healthy. My biggest joy is seeing people get healthier, stay healthy and slow the aging process as much as possible so they are vibrant, full of life and able to accomplish whatever brings them a sense of purpose.

CORE FOCUS ON RATIOS, RATES & STATISTICS
BLOCK 2 ~ RATES

LESSON 2.1	FRACTIONS AND DECIMALS	35
	EXPLORE! BACK AND FORTH	
LESSON 2.2	REPEATING DECIMALS AND ROUNDING	39
	EXPLORE! CALCULATORS AND FRACTIONS	
LESSON 2.3	RATES AND UNIT RATES	44
LESSON 2.4	RATE PROBLEM SOLVING	48
	EXPLORE! MATCH THE RATES	
LESSON 2.5	COMPARING RATES	54
	EXPLORE! SHOPPING SALES	
LESSON 2.6	MOTION RATES	60
REVIEW	BLOCK 2 ~ RATES	66

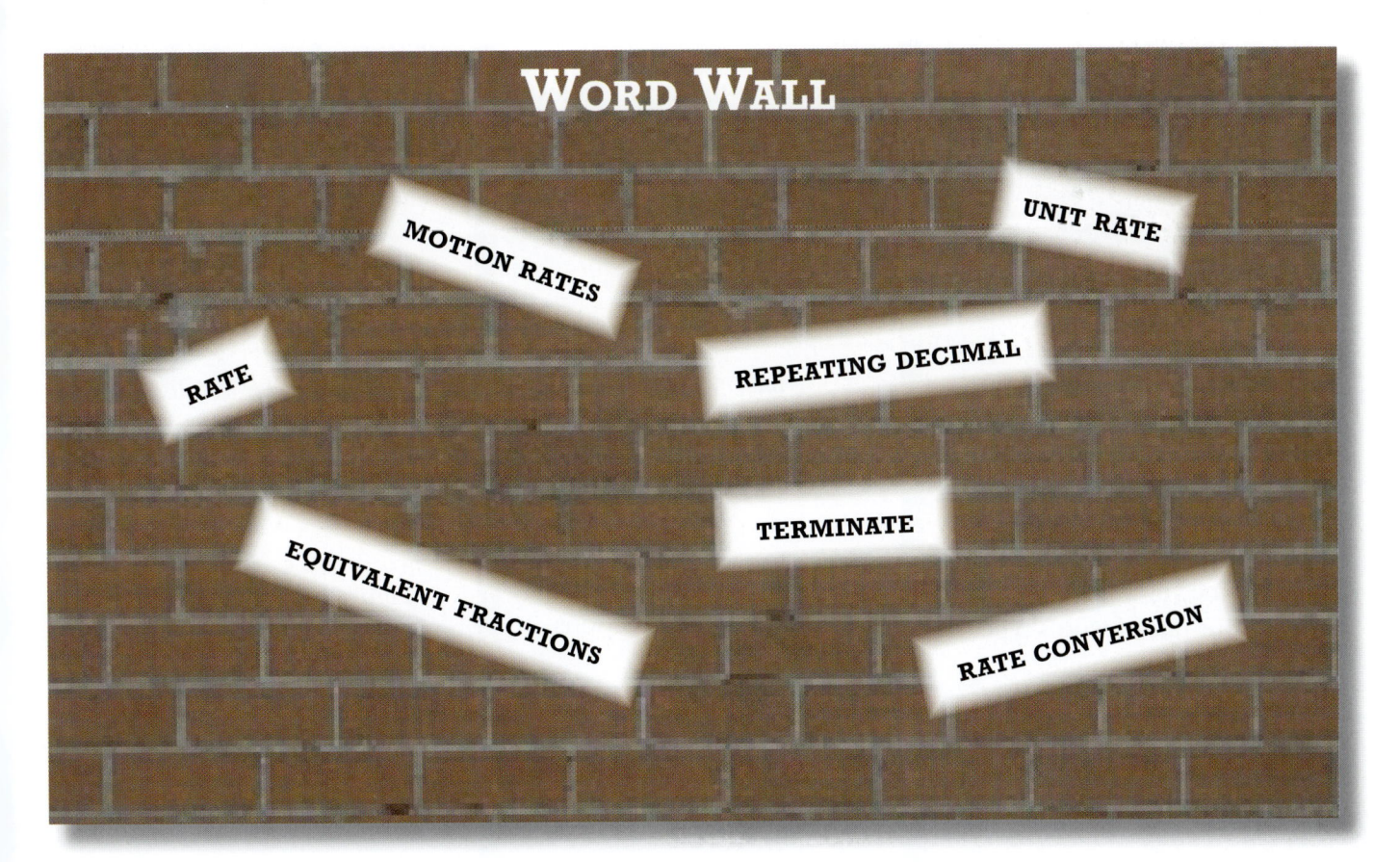

WORD WALL

- MOTION RATES
- UNIT RATE
- RATE
- REPEATING DECIMAL
- TERMINATE
- EQUIVALENT FRACTIONS
- RATE CONVERSION

BLOCK 2 ~ RATES
TIC-TAC-TOE

GAS MILEAGE Find the gas mileage of your family car and what the manufacturer says the gas mileage should be. *See page 52 for details.*	**CHILDREN'S STORY** Write a children's story using three different rates that need to be converted. *See page 59 for details.*	**MOTION RATE** Find the time, in miles per hour, it takes you to walk and run 0.25 miles. Use this rate to answer questions. *See page 65 for details.*
FOOD DILEMMA Take a trip to the grocery store. Find the best deal on cereal, peanut butter and cheese. *See page 58 for details.*	**DOES SPEEDING HELP?** Find the amount of time it takes a car to travel 2 miles in a construction zone at different speeds. *See page 59 for details.*	**BATTING AVERAGES** Calculate batting averages. Research batting averages in Major League Baseball. *See page 43 for details.*
TYPING Time your typing to find your rate of words typed per minute. *See page 53 for details.*	**FRACTIONS AND DECIMALS GAME** Create a matching game with equivalent fractions and decimals. *See page 65 for details.*	**BANKING** Figure out how much banks earn by rounding up or down on statements. *See page 47 for details.*

FRACTIONS AND DECIMALS

LESSON 2.1

Convert fractions to decimals and decimals to fractions.

Jacob ran $\frac{1}{2}$ mile and Sam ran $\frac{2}{5}$ mile in four minutes. Who ran farther during the four minutes?

One method of comparing fractions is finding a common denominator and then comparing the numerators to determine which fraction is greater. Another method is rewriting each fraction as a decimal and comparing the decimals.

The fraction bar is a way of showing division. This means $\frac{1}{2}$ can also be written $1 \div 2$. To write $\frac{1}{2}$ as a decimal find the value of $1 \div 2$.

$$2 \overline{)1.0} = 0.5 \qquad \frac{1}{2} = 0.5$$

To write $\frac{2}{5}$ as a decimal find the value of $2 \div 5$.

$$5 \overline{)2.0} = 0.4 \qquad \frac{2}{5} = 0.4$$

Since 0.5 is larger than 0.4 it can be written as $0.5 > 0.4$. This means $\frac{1}{2}$ is larger than $\frac{2}{5}$, or $\frac{1}{2} > \frac{2}{5}$. Since Jacob ran $\frac{1}{2}$ mile during the four minutes, Jacob ran farther than Sam.

EXPLORE!   BACK AND FORTH

Each tick mark between inches on a customary ruler represents one-sixteenth of an inch. They can be simplified to the following fractions:

$$\frac{1}{16}, \frac{1}{8}, \frac{3}{16}, \frac{1}{4}, \frac{5}{16}, \frac{3}{8}, \frac{7}{16}, \frac{1}{2}, \frac{9}{16}, \frac{5}{8}, \frac{11}{16}, \frac{3}{4}, \frac{13}{16}, \frac{7}{8}, \frac{15}{16}, 1$$

Step 1: Copy the table. Use a calculator to find the correct decimal value for each fraction. Complete the table.

Fraction	$\frac{1}{16}$	$\frac{1}{8}$	$\frac{3}{16}$	$\frac{1}{4}$	$\frac{5}{16}$	$\frac{3}{8}$	$\frac{7}{16}$	$\frac{1}{2}$	$\frac{9}{16}$	$\frac{5}{8}$	$\frac{11}{16}$	$\frac{3}{4}$	$\frac{13}{16}$	$\frac{7}{8}$	$\frac{15}{16}$	1
Decimal																

Step 2: Carla measured the length of a piece of wood. It was $8\frac{1}{4}$ inches long. Rewrite this measurement as a decimal using the table. Explain how the table was useful.

Step 3: Pam measured the length of a different piece of wood. It was 10.625 inches long. Rewrite this measurement as a mixed number using the table.

EXPLORE! CONTINUED

Step 4: Use the table in **Step 1** to write each decimal as a fraction or mixed number.
- a. 0.625
- b. 2.25
- c. 5.8125
- d. 3.5
- e. 12.75
- f. 9.1875

Sometimes it is helpful to write decimals as fractions. The place value of the last number in the decimal tells you which number to put in the denominator.

1000	100	10	1	0.1	0.01	0.001
Thousands	Hundreds	Tens	Ones	Tenths	Hundredths	Thousandths
			0	1	6	

Whole Number — Less Than One

For example, the decimal in the place value chart above is read "sixteen hundredths". It can be written as a fraction.

$$0.16 = \frac{16}{100} = \frac{4}{25}$$

WRITING A FRACTION AS A DECIMAL
1. Divide the numerator by the denominator.

WRITING A DECIMAL AS A FRACTION
1. Determine the place value of the last number in the decimal.
2. Write a fraction using the decimal as the numerator and the place value as the denominator.
3. Write the fraction in simplest form.

EXAMPLE 1

Convert each decimal to a fraction or mixed number in simplest form.
- a. 0.7
- b. 0.25
- c. 6.2

SOLUTIONS

a. Write 0.7 in words.
Use 10 as the denominator.

0.7 = seven tenths
$\frac{7}{10}$

b. Write 0.25 in words.
Use 100 as the denominator.
Write in simplest form.

0.25 = twenty-five hundredths
$\frac{25}{100}$
$\frac{1}{4}$

c. Write 6.2 in words.
Use 10 as the denominator.
Write in simplest form.

6.2 = six and two tenths
$6\frac{2}{10}$
$6\frac{1}{5}$

Lesson 2.1 ~ Fractions and Decimals

EXERCISES

Convert each fraction or mixed number to a decimal.

1. $\frac{2}{5}$

2. $\frac{1}{8}$

3. $\frac{7}{10}$

4. $\frac{1}{16}$

5. $\frac{5}{8}$

6. $\frac{1}{4}$

7. $2\frac{1}{2}$

8. $1\frac{3}{4}$

9. $10\frac{3}{5}$

10. Micaela measured the height of a candle in her room. It was $7\frac{1}{4}$ inches tall.
    a. Write the height of the candle as a decimal.
    b. Micaela measured the candle again after burning it. It was $4\frac{1}{8}$ inches tall. How much of the candle burned? Write your answer as a decimal and show all work necessary to justify your answer.

11. Tricia and Natalia converted $\frac{5}{8}$ to decimal form. Which person did the problem correctly? Explain the mistake in the other person's work.

Tricia's Work	Natalia's Work
$\frac{5}{8} = 8 \div 5$	$\frac{5}{8} = 5 \div 8$
$\begin{array}{r} 1.6 \\ 5\overline{)8.0} \\ \underline{-5} \\ 30 \\ \underline{-30} \\ 0 \end{array}$	$\begin{array}{r} 0.625 \\ 8\overline{)5.000} \\ \underline{-48} \\ 20 \\ \underline{-16} \\ 40 \\ \underline{-40} \\ 0 \end{array}$
$\frac{5}{8} = 1.6$	$\frac{5}{8} = 0.625$

Determine which fraction is larger by first rewriting the fractions as decimals to compare them.

12. $\frac{5}{8}$ and $\frac{1}{2}$

13. $\frac{3}{16}$ and $\frac{1}{4}$

14. $\frac{4}{5}$ and $\frac{7}{8}$

Write each decimal as a fraction or mixed number in simplest form.

15. 0.3

16. 0.6

17. 0.5

18. 0.25

19. 0.15

20. 1.25

21. 9.375

22. 10.2

23. 4.02

Write a fraction in simplest form that represents each model. Convert each fraction to a decimal.

24.

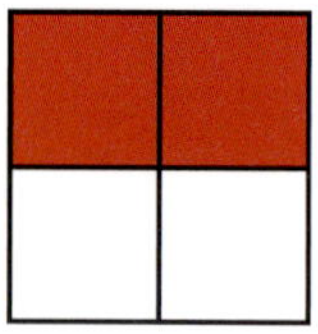

25.

26.

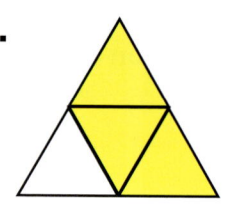

27. Each tick mark on a ruler between centimeters represents one millimeter or 0.1 centimeters. The table shows the value of each tick mark between 0 and 1 centimeter on a ruler in decimal form.

Decimal	0.1	0.2	0.3	0.4	0.5	0.6	0.7	0.8	0.9	1.0
Fraction										

  a. Copy and complete the table. Convert each decimal to a fraction in simplest form.
  b. Do you think it is easier to write parts of centimeters as decimals or fractions? Explain your reasoning.
  c. Pedro measured a piece of yarn. It was $7\frac{4}{5}$ centimeters long. Write this measurement as a decimal.

28. Which of the numbers below are equivalent to $4\frac{1}{4}$? Write all of the numbers that apply.
  4.25   4.14 $\frac{17}{4}$ $4\frac{2}{8}$ $4\frac{1}{5}$

29. Paul walked 3.7 miles. Siri walked $3\frac{4}{5}$ miles. David walked $3\frac{3}{4}$ miles. Which person walked the farthest? Show all work necessary to justify your answer.

REVIEW

Find the ratio of each geometric sequence. Use the ratio to find the next two terms of the geometric sequence.

30. 1, 3, 9, 27, 81, …

31. 4, 20, 100, 500, …

32. 100, 10, 1, 0.1, …

Complete each conversion. Show all work necessary to justify your answer.

33. 10 yards = _____ feet

34. 18 inches = _____ feet

35. 2,000 meters = _____ kilometers

36. 14 meters = _____ centimeters

37. Patrick walked 2.5 kilometers to work today. Convert this distance to meters.

REPEATING DECIMALS AND ROUNDING

LESSON 2.2

Convert fractions to repeating decimals.
Round numbers to a given place value.

Marcus' teacher asked him to write $\frac{1}{3}$ as a decimal.

He began by finding 1 ÷ 3.

$$3 \overline{)1.0000...} \quad 0.3333...$$

"This will keep going forever!" he realized. He wondered what to do.

Sometimes when you divide the numerator of a fraction by its denominator, the decimal does not **terminate** or stop. Instead, it keeps going. If a decimal has one or more digits that repeat forever, it is a **repeating decimal**.

When a decimal is a repeating decimal, write the repeating pattern once and draw a bar above the repeating part of the decimal. For example, Marcus would show that the 3 continues forever by writing:

$$\frac{1}{3} = 0.\overline{3}$$

EXAMPLE 1 Convert each fraction to a decimal.

a. $\frac{2}{3}$ b. $\frac{1}{6}$ c. $\frac{1}{11}$

SOLUTIONS

a. $\frac{2}{3} = 2 \div 3 = $ $3\overline{)2.0000...}$ 0.6666... $\frac{2}{3} = 0.\overline{6}$

The bar is only above the 6 because it is the only number that repeats.

b. $\frac{1}{6} = 1 \div 6 = $ $6\overline{)1.0000...}$ 0.1666... $\frac{1}{6} = 0.1\overline{6}$

The bar is above the 09 because both numbers repeat.

c. $\frac{1}{11} = 1 \div 11 = $ $11\overline{)1.0000...}$ 0.0909... $\frac{1}{11} = 0.\overline{09}$

Lesson 2.2 ~ Repeating Decimals and Rounding

EXPLORE!   CALCULATORS AND FRACTIONS

Step 1: Convert each fraction to a decimal.

$$\frac{1}{3} \qquad \frac{1}{9} \qquad \frac{8}{9} \qquad \frac{5}{6}$$

Step 2: Use a calculator to write each fraction in **Step 1** as a decimal. Divide the numerator of the fraction by its denominator. Write all numbers on the screen of the calculator as the answer.

Step 3: Are the answers in **Step 2** different than the answers in **Step 1**? Explain why or why not.

Step 4: Samantha says $\frac{2}{3} = 0.666667$. Pam says $\frac{2}{3} = 0.\overline{6}$. Who is correct? Explain your reasoning.

A calculator has limited space on the display screen. Because it cannot show *all* the numbers in a repeating decimal, it rounds the last digit on the screen.

You will often round decimal solutions. Use place value to round to the appropriate number.

1000	100	10	1	0.1	0.01	0.001
Thousands	Hundreds	Tens	Ones	Tenths	Hundredths	Thousandths

Whole Number · Less Than One

EXAMPLE 2 Round $2.\overline{4}$ to the nearest hundredth.

SOLUTION
Underline the number in the hundredths place. 2.4<u>4</u>4...

Look at the digit one place to its right. 2.44<u>4</u>...

Round down since 4 is less than 5. $2.\overline{4} \approx 2.44$

ROUNDING

Look at the digit one place to the right of the place value you need to round. If the digit is:
- 0, 1, 2, 3 or 4, round the number down so the digit in the place value you are rounding stays the same.
- 5, 6, 7, 8 or 9, round the number up one digit in the place value to which you are rounding.

Lesson 2.2 ~ Repeating Decimals and Rounding

EXAMPLE 3 Convert $\frac{2}{3}$ to a decimal rounded to the nearest hundredth.

SOLUTION

Convert $\frac{2}{3}$ to a decimal.	$\frac{2}{3} = 0.666...$
Underline the number in the hundredths place.	$0.6\underline{6}66....$
Look at the digit one place to its right.	$0.6\underline{6}66....$
Round up since 6 is more than 5.	$0.666... \approx 0.67$

$\frac{2}{3} \approx 0.67$

Notice that after rounding a decimal number, the rest of the digits are dropped.

EXERCISES

Convert each fraction to a decimal. If it is a repeating decimal, use a bar to show which number(s) repeat.

1. $\frac{2}{3}$
2. $\frac{2}{9}$
3. $\frac{7}{9}$
4. $\frac{5}{11}$
5. $\frac{3}{4}$
6. $\frac{1}{6}$
7. $\frac{1}{11}$
8. $\frac{1}{3}$
9. $\frac{4}{9}$

10. Juan had a piece of fabric $\frac{1}{4}$ yard long. Lucinda had a piece of fabric $\frac{3}{4}$ yard long. They wanted to know how much fabric they had combined.
    a. Juan added the fractions together to find the sum. Find the sum like Juan did.
    b. Lucinda converted each fraction to a decimal. She added the decimals to find the sum. Convert the fractions to decimals and find the sum like Lucinda did.
    c. Should Juan and Lucinda have the same answer? Do you have the same answer in **part a** as in **part b**? Explain your reasoning.

11. Justin had a piece of wire $\frac{1}{3}$ meter long. Sherry had a piece of wire $\frac{2}{3}$ meter long. They wanted to know how much wire they had all together.
    a. Justin added the fractions together to find the sum. Find the sum like Justin did.
    b. Sherry converted each fraction to a decimal. She added the decimals to find the sum. Convert the fractions to decimals and find the sum like Sherry did.
    c. Should Justin and Sherry have the same answer? Do you have the same answer in **part a** as in **part b**? Explain your reasoning.

Round each number to the nearest tenth.

12. $0.\overline{17}$

13. $0.\overline{3}$

14. $4.\overline{48}$

15. $10.\overline{26}$

Round each number to the nearest hundredth.

16. $0.\overline{3}$

17. $5.\overline{07}$

18. $0.\overline{8}$

19. $11.2\overline{13}$

Round each number to the nearest thousandth.

20. $0.1\overline{45}$

21. $23.\overline{4}$

22. $0.1\overline{09}$

23. $0.\overline{285714}$

Convert each fraction to a decimal. Round the answer to the nearest hundredth.

24. $\frac{2}{3}$

25. $\frac{7}{9}$

26. $\frac{2}{11}$

27. $\frac{5}{6}$

28. The fraction $\frac{22}{7}$ is often used in a formula to estimate the area of a circle.
   a. Write $\frac{22}{7}$ as a decimal.
   b. Does the decimal for $\frac{22}{7}$ terminate or repeat?
   c. Round the decimal for $\frac{22}{7}$ to the nearest hundredth.

29. Marci and Adalya ran a 100 meter dash. Marci ran the distance in 12.025 seconds. It took Adalya 12.031 seconds.
   a. Which runner had the faster time?
   b. The timers' stop watches rounded the times to the nearest hundredth. Would you be able to tell who won the race based on the stop watch times? Explain your reasoning.

30. Julius spent $2.94 on a music download. Carl spent $2.89 on a music download. Laura spent $3.05 on a music download. Sarah says they all paid about the same amount. Jon says only two of them paid about the same amount. Explain how both Sarah and Jon can be correct.

31. Nicolas was training for a race. In the actual race, times are recorded to the nearest hundredth. Nicolas only has a stopwatch that rounds times to the nearest tenth. After he ran the race, the time on his stopwatch showed 3.2 minutes. To the nearest hundredth, what is the longest amount of time it might have taken him to run the race? Explain how you know your answer is correct.

42 Lesson 2.2 ~ Repeating Decimals and Rounding

REVIEW

Convert each decimal to a fraction in simplest form.

32. 0.2

33. 0.75

34. 0.375

Determine which fraction is larger. Convert each fraction to a decimal. Compare each decimal.

35. $\frac{1}{3}$ or $\frac{3}{10}$

36. $\frac{2}{3}$ or $\frac{3}{4}$

37. $\frac{2}{5}$ or $\frac{1}{2}$

Tic-Tac-Toe ~ Batting Averages

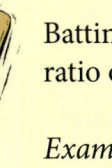

Batting averages for baseball and softball players are computed by finding the ratio of the number of hits a batter has to his/her number of times at bat.

Example: During the season Alex has batted 64 times. He has 19 hits. His batting average is:

$$\text{Batting Average} = \frac{\text{number of hits}}{\text{number of at bats}} = \frac{19}{64}$$

Although $\frac{19}{64}$ is the ratio describing Alex's batting average, batting averages are always expressed as decimals rounded to the nearest thousandths place. Alex's batting average = $\frac{19}{64}$ = 0.296875 ≈ 0.297. This is read as, "Alex's batting average is 297."

1. Use a calculator to find the batting average for each player on the school baseball team after 10 games.

Player	Jones	Field	Gonzales	Nguyen	Huff	Smith	Kent	Gwynn	Raxter	Brady
Hits	12	11	14	7	12	15	2	16	13	13
At Bats	40	36	42	35	38	40	12	40	34	41
Batting Average										

2. Which player had the highest batting average?

3. If Kent had 8 more at bats and 4 more hits, what would his new batting average be?

4. Ted Williams once had a batting average above 0.400 (read "400") at the end of a Major League Baseball season. Since then, other players have tried to hit that high of an average but no one has. Baseball is said to be a sport with many failures; you fail to get a hit more often than you succeed. Find the players with the top batting averages in both the National League and the American League for the past three years. Record the information. What is the highest batting average any player had in the last three years?

Lesson 2.2 ~ Repeating Decimals and Rounding

RATES AND UNIT RATES

LESSON 2.3

Convert rates to unit rates.

Dea found two different deals online for downloading songs onto her computer. Songs Now charges $4.60 for 20 songs. Let's Sing charges $0.25 per song. She wanted to figure out which company charges less money per song.

Dea will compare rates. A **rate** is a comparison of two numbers with different units. In this case, she will compare the units of dollars and songs. Songs Now charges a rate of $\frac{\$4.60}{20 \text{ songs}}$. Let's Sing charges a rate of $\frac{\$0.25}{1 \text{ song}}$.

Let's Sing's rate is a special rate called a unit rate. A **unit rate** is a rate that can be written as a fraction with a denominator of 1. These rates can also be written as a single number using the word *per* or using a fraction bar to explain the units.

$$\frac{\$0.25}{1 \text{ song}} = \$0.25 \text{ per song} = 25 \text{ cents/song}$$

Write the rate for Songs Now as a unit rate. To compare the prices of the two companies, the rates should be written as unit rates.

Rewrite the rate $\frac{\$4.60}{20 \text{ songs}}$ so it has a denominator of 1.

Divide the numerator and denominator by 20 so the new denominator will be 1.

$$\frac{\$4.60}{20 \text{ songs}} \xrightarrow{\div 20} \frac{\$0.23}{1 \text{ song}}$$

This means Songs Now charges $0.23 per song or 23 cents/song.

Songs Now charges less per song than Let's Sing since $0.23 per song is less than $0.25 per song. Dea chose to buy songs from Songs Now.

EXAMPLE 1 Find each unit rate.

a. $\frac{50 \text{ miles}}{2 \text{ hours}}$ b. $\frac{\$2.40}{3 \text{ candy bars}}$

SOLUTIONS

a. Rewrite the fraction with a denominator of 1.
The unit rate is 25 miles per hour.

$$\frac{50 \text{ miles}}{2 \text{ hours}} \xrightarrow{\div 2} \frac{25 \text{ miles}}{1 \text{ hour}}$$

b. Rewrite the fraction with a denominator of 1.
The unit rate is $0.80 per candy bar.

$$\frac{\$2.40}{3 \text{ candy bars}} \xrightarrow{\div 3} \frac{\$0.80}{1 \text{ candy bar}}$$

Finding a unit rate is similar to rewriting a fraction as a decimal. One thing that is different is that a unit rate includes the labels for the units.

EXAMPLE 2   The United States Mint in Philadelphia produces 2,250 coins every 30 minutes. Find how many coins per minute the Mint produces.

SOLUTION

Write the ratio of the number of coins per 30 minutes as a rate.

$$\frac{2250 \text{ coins}}{30 \text{ minutes}}$$

Rewrite the fraction so it has a denominator of 1.

$$\frac{2250 \text{ coins}}{30 \text{ minutes}} \xrightarrow{\div 30} = \frac{75 \text{ coins}}{1 \text{ minute}} \xleftarrow{\div 30}$$

The Mint produces 75 coins per minute.

EXERCISES

Find each unit rate.

1. $\dfrac{60 \text{ miles}}{2 \text{ hours}}$

2. $\dfrac{96 \text{ words}}{4 \text{ minutes}}$

3. $\dfrac{300 \text{ miles}}{15 \text{ gallons}}$

4. $\dfrac{12 \text{ days}}{2 \text{ jobs}}$

5. $\dfrac{\$3.30}{3 \text{ pens}}$

6. $\dfrac{24 \text{ ounces}}{1.5 \text{ servings}}$

7. $\dfrac{32 \text{ pounds}}{8 \text{ inches}}$

8. $\dfrac{30 \text{ feet}}{3 \text{ seconds}}$

9. $\dfrac{15 \text{ kilometers}}{5 \text{ hours}}$

10. Jaden and Jaxen knew they could skateboard 3 miles in 30 minutes. They figured out their speed in miles per minute. Their work is below.

Jaden	Jaxen
$\dfrac{3 \text{ miles} \div 30}{30 \text{ minutes} \div 30} = \dfrac{0.1 \text{ miles}}{1 \text{ minute}}$	$\dfrac{3 \text{ miles}}{30 \text{ minutes}} = 30 \overline{)3.0}^{\,0.1}$
0.1 miles per minute	0.1 miles per minute

a. Explain how Jaden solved the problem.
b. Explain how Jaxen solved the problem.
c. What number did both boys divide by to get their answers?
d. Their friend, Max, wanted to find their rate in miles per hour. His calculations are shown below. He said Jaden and Jaxen skateboarded 6 miles per hour. Is this correct? Explain your reasoning.

$$\frac{3 \text{ miles} \div 0.5}{0.5 \text{ hour} \div 0.5} = \frac{6 \text{ miles}}{1 \text{ hour}}$$

Lesson 2.3 ~ Rates and Unit Rates

Show all work necessary to justify your answer for each Exercise.

11. José spent $36 for 4 movie tickets. Find the price per ticket.

12. Rob spent $3.30 for 6 large cookies. Find the price per cookie.

13. Polly used 10 gallons of gas to drive 235 miles on a trip. Find how many miles per gallon Polly's car got on the trip.

14. Tran rode his scooter 10 miles in 1.5 hours. Find how many miles per hour he rode.

15. Maria could buy 6 songs online for $3.00 at Songs-R-Us, or she could pay $0.45 per song at Music Hooray. Which company charges less per song? Use mathematics to justify your answer.

16. Luke walked 2 miles in 40 minutes. He determined his unit rate was 20 minutes per mile. Sally informed him his rate was 0.05 miles per minute. Explain how both of these unit rates are accurate.

REVIEW

Convert each fraction to a decimal. If it is a repeating decimal, use a bar to show which number(s) repeat.

17. $\frac{1}{3}$ **18.** $\frac{1}{2}$

19. $\frac{3}{8}$ **20.** $\frac{2}{3}$

Convert each decimal to a fraction. Write in simplest form.

21. 0.25   **22.** 0.06

23. 1.3   **24.** 0.8

Round each decimal to the nearest tenth.

25. $0.\overline{3}$ **26.** $2.\overline{09}$

27. $5.8\overline{3}$ **28.** $0.\overline{7}$

29. The ratio of boys to girls is the same in three classrooms. Each ratio is shown in the table. Find the missing value. Show all work necessary to justify your answer.

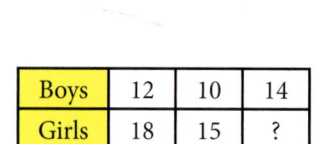

Boys	12	10	14
Girls	18	15	?

46   Lesson 2.3 ~ Rates and Unit Rates

Tic-Tac-Toe ~ Banking

A bank often rounds interest added to a savings account by rounding the value down to the nearest cent. This is called "truncating" the number.

Example: The interest added to Jim's savings account is $2.34976. Rather than rounding the number to the nearest penny ($2.35), the bank rounds down to $2.34 because they truncate the number 2.34|976. This means they ignore the digits after the hundredths place.

Step 1: Why would a bank round interest this way? Explain your reasoning using complete sentences.

Step 2: If 10,000 people save money in the bank and half of them have interest that should round up, about how much money will the bank save by rounding down to the nearest cent?

Step 3: If 100,000 people save money in the bank and half of them have interest that should round up, about how much money will the bank save by rounding down to the nearest cent?

A credit card company charges interest on items purchased using a credit card. The credit card company often rounds interest by rounding the value up to the nearest cent rather than truncating the number.

Example: The interest added to the credt card bill is $10.9813. The credit card company adds $10.99 to the bill even though the interest rounds to $10.98.

Step 4: Why would a credit card company round interest this way? Explain your reasoning using complete sentences.

Step 5: If 20,000 people use the credit card company and half of them have interest that should round down, about how much extra money will the credit card company get by rounding up to the nearest cent?

Step 6: If 1,000,000 people use the credit card company and half of them have interest that should round down, about how much extra money will the credit card company get by rounding up to the nearest cent?

Step 7: Most banks loan money and provide savings accounts for their customers. A national banking company had 10,000,000 customers with both a savings account and a credit account one month. Half of their customers had interest rounded up on their credit statements and down on their savings statements. How much extra money did the bank make that month? Show all work necessary to justify your answer.

RATE PROBLEM SOLVING

LESSON 2.4

Solve problems using equivalent rates and unit rates.

The fractions $\frac{2}{4}$, $\frac{3}{6}$, $\frac{4}{8}$, and $\frac{5}{10}$ can be written in simplest form as $\frac{1}{2}$. These are examples of **equivalent fractions**. Equivalent fractions are fractions with the same value. Since ratios and rates can be written as fractions, they also can be written in many equivalent forms.

EXPLORE! **MATCH THE RATES**

Below are 10 rates.

$$\frac{\$8.00}{4 \text{ tickets}} \qquad \frac{\$10.00}{2 \text{ tickets}} \qquad \frac{\$16.00}{2 \text{ tickets}}$$

$$\frac{\$12.00}{4 \text{ tickets}} \qquad\qquad\qquad\qquad \frac{\$5.00}{1 \text{ ticket}}$$

$$\frac{\$12.00}{6 \text{ tickets}} \qquad\qquad\qquad\qquad \frac{\$24.00}{4 \text{ tickets}}$$

$$\frac{\$6.00}{1 \text{ ticket}} \qquad \frac{\$9.00}{3 \text{ tickets}} \qquad \frac{\$24.00}{3 \text{ tickets}}$$

Step 1: Each rate has the same units. Write the units for the rates. (_____ per _____)

Step 2: Which of the above rates are already written as unit rates?

Step 3: There are five pairs of equivalent rates. One is given below. Find the four other pairs. Write the pairs next to one another with an equals sign between the two rates.

1. $\frac{\$12.00}{4 \text{ tickets}} = \frac{\$9.00}{3 \text{ tickets}}$ 2. 3. 4. 5.

Step 4: Explain how you figured out which rates were equivalent.

Step 5: The price for a ticket to a jazz concert was $14. Write 5 equivalent rates using the unit rate of $14 per ticket.

48 Lesson 2.4 ~ Rate Problem Solving

Problems involving rates can be solved using two different methods. You can use equivalent fractions or unit rates.

SOLVING PROBLEMS INVOLVING RATES

Using Equivalent Fractions
1. Write the two rates with an equals sign (=) between them.
2. Identify what you need to multiply the numerator or denominator of the complete rate by to write an equivalent rate on the other side of the equals sign.

Using Unit Rates
1. Find the unit rate for the known rate.
2. Multiply the unit rate by the known quantity.

EXAMPLE 1

Complete each equivalent rate.

a. $\dfrac{24 \text{ miles}}{1 \text{ gallon}} = \dfrac{\text{miles}}{6 \text{ gallons}}$

b. $\dfrac{\$6.00}{4 \text{ liters}} = \dfrac{\$}{32 \text{ liters}}$

SOLUTIONS

a. Find the factor from one denominator to the other.

$\dfrac{24 \text{ miles}}{1 \text{ gallon}} \xrightarrow{\times 6} = \dfrac{\text{miles}}{6 \text{ gallons}}$

Multiply the numerator by the same factor to complete the equivalent rate.

$\dfrac{24 \text{ miles}}{1 \text{ gallon}} \xrightarrow{\times 6} = \dfrac{144 \text{ miles}}{6 \text{ gallons}}$

b. Find the factor from one denominator to the other.

$\dfrac{\$6.00}{4 \text{ liters}} \xrightarrow{\times 8} = \dfrac{\$}{32 \text{ liters}}$

Multiply the numerator by the same factor to complete the equivalent rate.

$\dfrac{\$6.00}{4 \text{ liters}} \xrightarrow{\times 8} = \dfrac{\$48.00}{32 \text{ liters}}$

EXAMPLE 2

Nigel paid $3.60 for 30 copies of his flyer. Use a unit rate to determine the cost to make 80 copies of his flyer.

SOLUTION

Write the rate as a fraction.

$\dfrac{\$3.60}{30 \text{ copies}}$

Find the unit rate.

$\dfrac{\$3.60}{30 \text{ copies}} \xrightarrow{\div 30} = \dfrac{\$0.12}{1 \text{ copy}}$

Multiply the cost per copy by the number of copies.

$\$0.12 \times 80 = \$9.60$

Nigel will pay $9.60 to make 80 copies of his flyer.

Lesson 2.4 ~ Rate Problem Solving

EXAMPLE 3

Tom buys some apples at a local fruit stand. The fruit stand charges $3.00 for every 2 pounds. Find the price Tom pays for 12 pounds of apples.

SOLUTION

METHOD 1 ~ Equivalent Fractions
Write the rate as a fraction.

$$\frac{\$3.00}{2\ lbs}$$

Write a second fraction with a denominator of 12 pounds.

$$\frac{\$3.00}{2\ lbs} = \frac{\$}{12\ lbs}$$

The new denominator is 6 times the original denominator. Multiply the numerator by 6.

$$\frac{\$3.00}{2\ lbs} \xrightarrow{\times 6} \frac{\$18.00}{12\ lbs}$$

Tom pays $18 for 12 pounds of apples.

METHOD 2 ~ Unit Rates
Write the rate as a fraction.

$$\frac{\$3.00}{2\ lbs}$$

Find the unit rate.

$$\frac{\$3.00}{2\ lbs} \xrightarrow{\div 2} \frac{\$1.50}{1\ lb}$$

Multiply the cost per pound ($1.50) by the number of pounds.

$$\$1.50 \times 12 = \$18.00$$

Tom pays $18 for 12 pounds of apples.

EXERCISES

Complete each equivalent rate.

1. $\dfrac{\$3.00}{1\ gallon} = \dfrac{\$}{10\ gallons}$

2. $\dfrac{3\ miles}{1\ hour} = \dfrac{miles}{8\ hours}$

3. $\dfrac{60\ words}{2\ minutes} = \dfrac{words}{14\ minutes}$

4. $\dfrac{3\ kilometers}{1\ hour} = \dfrac{kilometers}{3\ hours}$

5. $\dfrac{25\ miles}{1\ gallon} = \dfrac{200\ miles}{gallons}$

6. $\dfrac{12\ jobs}{5\ days} = \dfrac{48\ jobs}{days}$

Use equivalent rates to complete each problem.

7. Felicia drove 120 miles in 3 hours. At this rate, how far will she drive in 6 hours?

8. Marcus burns 9 calories per minute when running. How long will he need to run to burn 270 calories? Show all work necessary to justify your answer.

9. Henry paid $60 for 5 people to attend a play on Broadway. Next month, 15 people in his class would like to go. If the cost is the same per ticket, how much will Henry pay for 15 people to attend next month?

Find each unit rate. Round to the nearest hundredth, if necessary.

10. $\dfrac{8 \text{ feet}}{2 \text{ minutes}}$

11. $\dfrac{\$4.00}{10 \text{ pencils}}$

12. $\dfrac{70 \text{ miles}}{3 \text{ gallons}}$

13. $\dfrac{12 \text{ meters}}{48 \text{ seconds}}$

14. $\dfrac{105 \text{ words}}{2 \text{ minutes}}$

15. $\dfrac{\$8.00}{12 \text{ books}}$

Use a unit rate to complete each problem.

16. Jimmy's new car went 204 miles using 12 gallons of gas. At this rate, how many miles can he travel using 5 gallons of gas? Show all work necessary to justify your answer.

17. Patrick went to the store to buy a seedless watermelon. It was on sale for $0.88 for every 2 pounds. He bought an 11 pound watermelon. How much did Patrick pay for the watermelon?

18. Denise filled her wading pool using her garden hose. The pool filled at a rate of 7 gallons every 2 minutes. She left the water on for 9 minutes. How many gallons of water were in the wading pool?

Use equivalent fractions or unit rates to solve each problem.

19. Aaron walked 8 miles in 2 hours. To determine how far he could walk at this rate in 6 hours, Aaron used a unit rate and showed the following work. Show another way to find how far Aaron can walk at that rate in 6 hours.

$$\dfrac{8 \text{ miles}}{2 \text{ hours}} = \dfrac{4 \text{ miles}}{1 \text{ hour}}$$

4 miles per hour × 6 hours = 24 miles

20. Josh spent $4.40 for 4 candy bars at the student store. How much would he pay for 7 candy bars at the student store?

21. Miranda's mom sent her to the grocery store with $20.00. She bought 2 pounds of roast beef, 3 pounds of apples, 1 loaf of bread and 1 gallon of milk. She could buy anything else at the store she wanted with the remaining money. Use the prices below to determine if she had enough money to purchase one cookie and one bag of popcorn. Show all work necessary to justify your answer.

Roast beef: $5.00 per pound
Apples: $2.50 for 2 pounds
Bread: $2.00 per loaf
Milk: $2.50 per gallon

Juice Box: $0.50 per box
Cookie: $1.50 for 2 cookies
Candy bar: $1.00 per candy bar
Popcorn: $1.25 per bag

Lesson 2.4 ~ Rate Problem Solving **51**

REVIEW

Complete each conversion. Show all work necessary to justify your answer.

22. 3 kilometers = _____ meters

23. 35 millimeters = _____ centimeters

24. 42 inches = _____ feet

25. 5 yards = _____ feet

26. Lonnie spent $9.30 on 3 small cakes. Find the price per cake.

27. Jeff walked 27 miles in 6 hours. Find his speed in miles per hour.

28. There were eight boys and 20 girls at a party when it started. When it ended, there were 12 boys at the party but the same ratio of boys to girls as when the party began. How many girls were at the party when it ended? Use mathematics to justify your answer.

TIC-TAC-TOE ~ GAS MILEAGE

The gas mileage a car gets is the ratio of the miles a car has driven to the number of gallons of gas used.

$$\text{Gas Mileage} = \frac{\text{miles driven}}{\text{gallons of gas used}}$$

Example: A car traveled 235 miles using 12 gallons of gas. Its gas mileage is $\frac{235 \text{ miles}}{12 \text{ gallons}}$. The rate for gas mileage is usually written as a decimal rounded to the nearest tenth.

In this case, $\frac{235 \text{ miles}}{12 \text{ gallons}} = 19.58\overline{3} \approx 19.6$ miles per gallon.

Step 1: Record the gas mileage of your family car. Do this by writing down the number of miles driven since the last fill-up and the amount of gas needed to fill up the tank at the gas station.

Step 2: Record the gas mileage of your family car one more time to compare the two rates.

Step 3: What is the estimated gas mileage for your family car based on your data?

Step 4: Research your car to find out what the manufacturer says the gas mileage should be.

Step 5: Research to find which cars have the best gas mileage (most miles per gallon). Create a list of the top five cars.

Tic-Tac-Toe ~ Typing

How many words per minute can you type? Use a timer or ask a friend to time you as you type the following story.

> Sally went to the store with her mother and brother and bought some milk, carrots, onions, salad dressing and tomatoes. Next, Sally's mom took her to the dentist and the dry cleaners. Sally wanted to go home and play with her friends. Finally, Sally's mom was done with errands for the day. She took Sally to the park to play with her friends. Sally's friend, Tom, asked her what she had done that day. She told Tom she went to the store, the dentist and the dry cleaners. Tom reminded her that they had soccer practice in the evening. Sally told him she would see him at practice. She left for home to get ready.

Step 1: Type the entire 115 word paragraph and time yourself. Record the number of seconds it took you to type the passage. Also record the number of errors you made. Keep typing the passage until you make fewer than 5 errors. If this happens on your first try, type faster and see how many errors you make. Type the passage and record the information at least three times.

Attempts	Time (*sec*)	Number of Errors
1		
2		
3		

Step 2: Convert the time it took you to type the passage from seconds to minutes. Round to the nearest hundredth.

Step 3: What was your fastest typing rate as a unit rate of words per minute?

Step 4: What was your fastest typing rate with fewer than 5 errors?

Step 5: How long would it take you to type a 1-page paper with 460 words at your fastest rate? Show all work necessary to justify your answer.

Lesson 2.4 ~ Rate Problem Solving

COMPARING RATES

LESSON 2.5

Compare rates using equivalent rates and unit rates.

EXPLORE!   **SHOPPING SALES**

Step 1: Kay went to the department store to buy new shirts for the 12 girls on her soccer team. She found a sale and could buy 3 shirts for $9.00.

   a. Write the rate as a fraction: $\dfrac{\$\_\_\_\_}{\text{shirts}}$

   b. Rewrite the rate as a unit rate. What is the price per shirt?

   c. How much will it cost Kay to buy 12 shirts at this price per shirt?

   d. Another way to find the cost is to use equivalent rates. Complete the equivalent rate to find Kay's cost for 12 shirts.

$$\frac{\$9.00}{3 \text{ shirts}} = \frac{\$\_\_\_\_}{12 \text{ shirts}}$$

Step 2: Trudy went to a different department store. There she could buy 4 shirts for $12.00.

   a. Write the rate as a fraction: $\dfrac{\$\_\_\_\_}{\text{shirts}}$

   b. Rewrite the rate as a unit rate. What is the price per shirt?

   c. How much will it cost Trudy to buy 12 shirts at this price per shirt?

   d. Another way to find the cost is to use equivalent rates. Complete the equivalent rate to find Trudy's cost for 12 shirts.

$$\frac{\$12.00}{4 \text{ shirts}} = \frac{\$\_\_\_\_}{12 \text{ shirts}}$$

Step 3: Mark can buy 2 pairs of jeans for $48.00 at Store A or 3 pairs of jeans for $66.00 at Store B. At which store will Mark pay less per pair of jeans? Explain how you know your answer is correct.

TO COMPARE RATES

1. Write each rate as a unit rate with the same units in the numerator and the denominator.
2. Compare the values of the numerators of the unit rates to determine which rate is larger or smaller.

EXAMPLE 1 | **Is it a better deal to buy a 20-ounce box of cereal for $3.50 or a 16-ounce box of cereal for $3.00?**

SOLUTION | To find the best deal, write each rate as a unit rate by finding dollars per ounce.

20-ounce box for $3.50 $\dfrac{\$3.50}{20\ oz} = \dfrac{\$0.175}{1\ oz}$ (÷ 20) $0.175 per ounce

16-ounce box for $3.00 $\dfrac{\$3.00}{16\ oz} = \dfrac{\$0.1875}{1\ oz}$ (÷ 16) $0.1875 per ounce

The 20-ounce box is the better deal because it costs less per ounce.

EXAMPLE 2 | **Which vehicle gets better gas mileage: a car that travels 408 miles using 12 gallons of gas or a truck that travels 448 miles using 14 gallons of gas?**

SOLUTION | To find the better gas mileage, write each rate as a unit rate by finding miles per gallon.

Car: $\dfrac{408\ \text{miles}}{12\ \text{gallons}} = \dfrac{34\ \text{miles}}{1\ \text{gallon}}$ (÷ 12)   34 miles per gallon

Truck: $\dfrac{448\ \text{miles}}{14\ \text{gallons}} = \dfrac{32\ \text{miles}}{1\ \text{gallon}}$ (÷ 14)   32 miles per gallon

The car gets better gas mileage because it travels more miles per gallon.

Lesson 2.5 ~ Comparing Rates

EXAMPLE 3

Paul went to the grocery store to buy potatoes. A 10-pound bag of potatoes cost $4.00. A 6-pound bag of potatoes cost $2.70.
a. Which size of bag is the best deal?
b. What is the lowest total cost Paul will pay for 20 pounds of potatoes?

SOLUTIONS

a. Compare the unit rates.

$4.00 for 10 *lbs*: $\dfrac{\$4.00}{10\ lbs} = \dfrac{\$0.40}{1\ lb}$ $0.40 per pound

(÷ 10)

$2.70 for 6 *lbs*: $\dfrac{\$2.70}{6\ lbs} = \dfrac{\$0.45}{1\ lb}$ $0.45 per pound

(÷ 6)

It is cheaper per pound to buy the 10-pound bag.

b. Use equivalent rates.

$4.00 for 10 *lbs*: $\dfrac{\$4.00}{10\ lbs} = \dfrac{\$}{20\ lbs}$

$\dfrac{\$4.00}{10\ lbs} = \dfrac{\$8.00}{20\ lbs}$ (× 2)

Twenty pounds of potatoes cost $8.00.

EXERCISES

Use unit rates to determine which of the two rates is smaller.

1. $\dfrac{\$5.00}{2\ \text{sandwiches}}$ or $\dfrac{\$6.75}{3\ \text{sandwiches}}$

2. $\dfrac{10\ \text{miles}}{1\ \text{hour}}$ or $\dfrac{35\ \text{miles}}{4\ \text{hours}}$

3. $\dfrac{162\ \text{miles}}{6\ \text{gallons}}$ or $\dfrac{150\ \text{miles}}{5\ \text{gallons}}$

4. $\dfrac{16\ \text{jobs}}{4\ \text{days}}$ or $\dfrac{10\ \text{jobs}}{2\ \text{days}}$

5. Kyle types at a rate of 55 words per minute. Christine types at a rate of 120 words in 2 minutes. Which person types more words per minute? Explain your reasoning.

6. Marta runs a race against her best friend Markesha. Marta runs at a rate of 7 miles in 1 hour. Markesha runs at a rate of 4 miles in 0.5 hours. Which person runs at a faster rate? Use mathematics to justify your reasoning.

7. Ivan needs new highlighters. He can buy a package of 6 highlighters for $7.50 or a package of 4 highlighters for $6.00. Which package of highlighters has the best price per highlighter?

8. Lyuba needs to buy dog food for her dog. She can buy a 20-pound bag of dog food for $15.00 or a 40-pound bag for $28.00. Which bag of dog food is cheaper per pound? Show all work necessary to justify your answer.

9. Mark drove 150 miles in 3 hours. Jamal drove 220 miles in 4 hours.
   a. Who drove faster? Show all work necessary to justify your answer.
   b. Both men traveled at these rates for a total of 6 hours. How far did each one travel?

Use equivalent rates to determine the total cost for 24 pounds of marionberries from three local farms given the price each farm charges.

10. $1.50 per pound

11. $5.00 for 3 pounds

12. $4.00 for 2 pounds

13. Ryan needs to buy 60 notebooks. At the store he found he could buy a package of 20 notebooks for $20.00 or a package of 15 notebooks for $10.00.
   a. How many 20-notebook packages would Ryan need to buy?
   b. How much will Ryan pay if he chooses to buy packages of 20?
   c. How much will Ryan pay if he chooses to buy packages of 15?
   d. Which packages should Ryan buy if he wants the cheapest cost?
   e. Check your answer. Find the unit rate (price per notebook) for the 20-notebook package and the 15-notebook package.

14. Shelly rides her 10-speed bike at a rate of 16 miles per hour. She rides her mountain bike 24 miles in 2 hours. She needs to ride 48 miles. Which bike should she ride to get there most quickly?
   a. Find the unit rate of speed (miles per hour) for the 10-speed.
   b. How long would it take Shelly to ride the 10-speed 48 miles?
   c. How long would it take Shelly to ride the mountain bike 48 miles?
   d. On which bike will Shelly ride the 48 miles faster? Explain your reasoning.

15. Zane needs to buy 20 balloons for a birthday party. He can buy a package of 5 balloons for $4.00 or a package of 4 balloons for $3.60. How much will it cost him to buy 20 of the cheaper balloons? Show all work necessary to justify your answer.

REVIEW

Convert each fraction or mixed number to a decimal. If it is a repeating decimal, use a bar to show which number(s) repeat.

16. $\frac{3}{4}$

17. $\frac{2}{3}$

18. $2\frac{4}{5}$

19. $1\frac{1}{2}$

20. Three pigs and 9 goats live at a local farm.
  a. Write the ratio of pigs to goats.
  b. Write the ratio of pigs to animals at the farm.
  c. Write the ratio of goats to animals at the farm.

21. Six out of 9 boys surveyed like soccer.
  a. Write the ratio of boys who like soccer to boys surveyed.
  b. Write the ratio of boys who like soccer to boys who do not like soccer.

Tic-Tac-Toe ~ Food Dilemma

Grocery stores sell the same type of cereal in different-sized boxes. There is one price for a 14-ounce box and another price for a 20-ounce box of the same brand. Which one is the best deal?

Take a trip to a grocery store to find the items listed in the chart. Remember you DO NOT have to buy the items, just find the prices.

Step 1: Copy the following table and take it to a local grocery store.

Step 2: Record the brand name of the item you have selected for cereal, cheese and peanut butter.

Step 3: Record the size of the item and its price. If it is on sale, put a * next to the price.

Step 4: Find the unit price for each item (the price per ounce or price per pound).

Step 5: Determine which size item is the best deal (cheapest price per ounce or pound).

	Size	Price	Unit Price	Best Deal?
Cereal Brand: _____	1. 2. 3.	1. 2. 3.	1. 2. 3.	
Cheese Brand: _____	1. 2. 3.	1. 2. 3.	1. 2. 3.	
Peanut Butter Brand: _____	1. 2. 3.	1. 2. 3.	1. 2. 3.	

Tic-Tac-Toe ~ Does Speeding Help?

There is a two-mile stretch of highway under construction. The speed limit in the construction zone is 20 miles per hour. The fines for speeding on that section of the road begin at $280 and increase according to the speed of the driver. If a person chooses to speed in the construction zone, how much time will they really save?

Step 1: Find the amount of time it takes to drive two miles at 20 miles per hour. You need to know the time it takes to drive 1 mile to find the time for 2 miles. Change the ratio from 20 miles per hour to 1 hour for 20 miles and find the unit rate.

$$\frac{1 \text{ hour}}{20 \text{ miles}} = \frac{? \text{ hour}}{1 \text{ mile}}$$

Convert the decimal for the number of hours to minutes → _____ minutes
How many minutes does it take a driver to drive the 2 miles at 20 miles per hour?

Step 2: Use the steps above to find the number of minutes it takes a driver to drive the 2 mile section of road at each of the speeds in the chart. Copy and complete the chart.

Miles per hour	15	20	25	30	40	50
Minutes to travel 2 miles						

Step 3: Explain why it is not better to drive 50 miles per hour than 20 miles per hour in the construction zone. Use complete sentences and include information from your table in your explanation.

Tic-Tac-Toe ~ Children's Story

Step 1: Create a children's book that incorporates three different rates that need to be converted. For example, convert miles per hour to feet per hour or jobs per week to jobs per day.

Step 2: The story may also involve comparison of rates. Look through this textbook to get more ideas about the three different rates to use in your story.

Step 3: Your book should have a cover, illustrations and an appropriate story line for children.

MOTION RATES

LESSON 2.6

Solve problems, compare and convert motion rates.

Rates that compare distance to time are called **motion rates**. Meters per hour, feet per hour and inches per minute are examples of motion rates. Distance can be measured using customary or metric units. Time is most often written as seconds, minutes, hours, days, weeks, months or years.

If you know the unit rate someone was traveling at and the amount of time they traveled, you can determine how far they traveled.

EXAMPLE 1

Oksana's family traveled on a highway at a rate of 60 miles per hour. Her family traveled at this rate for 2.5 hours. How far did they drive?

SOLUTION

Locate the unit rate.   60 miles per hour

Multiply by 2.5 hours.   60 × 2.5 = 150

Oksana's family traveled 150 miles in 2.5 hours.

You can use unit rates to compare two motion rates.

EXAMPLE 2

Wayne and Marla each left home on their bikes. They were meeting at the library, which is exactly the same distance from each of their homes. Wayne traveled at a rate of 20 miles every 2 hours. Marla traveled at a rate of 6 miles every 0.5 hour. They left home at the same time. Who arrived at the library first?

SOLUTION

Find Wayne's unit rate of speed. $\frac{20 \text{ miles}}{2 \text{ hours}} = \frac{10 \text{ miles}}{1 \text{ hour}}$ (÷ 2)

Wayne traveled 10 miles per hour.

Find Marla's unit rate of speed. $\frac{6 \text{ miles}}{0.5 \text{ hour}} = \frac{12 \text{ miles}}{1 \text{ hour}}$ (÷ 0.5)

Marla traveled 12 miles per hour.

Marla rode faster, so she arrived at the library first.

Martin traveled at a rate of 3 kilometers per hour on his tricycle. What is this unit rate when converted to meters per hour?

Karla walked at a rate of 4 miles per hour. What is this unit rate when converted to feet per hour?

Justin watched a bug travel at a rate of 3 feet per minute. How fast was the bug traveling when measured in feet per hour?

Each of these situations requires a **rate conversion**. A rate conversion is performed by changing at least one of the units in the rate. Three kilometers per hour can be changed to meters per hour by converting kilometers to meters.

Look at Martin's rate on his tricycle. Find an equivalent measurement relating kilometers and meters. Since 1 kilometer = 1000 meters, this will be used to make the conversion rate. Multiply the original rate by a conversion rate so unwanted units will cancel.

$$\frac{3 \text{ kilometers}}{1 \text{ hour}} \times \frac{1000 \text{ meters}}{1 \text{ kilometer}} = \frac{3000 \text{ meters}}{1 \text{ hour}}$$

Just like multiplication with numbers, common units can cancel if they are in the numerator and denominator. In this case, "kilometers" cancels.

original rate × conversion rate = new equivalent rate

Converting Rates
1. Write the rate as a fraction.
2. Identify the units you want in the answer.
3. Multiply the original rate by a conversion rate using equivalent measurements. Make sure the unwanted units cancel to get the answer.

EXAMPLE 3   Convert 4 miles per hour to feet per hour.

SOLUTION

Write the rate as a fraction. $\frac{4 \text{ miles}}{1 \text{ hour}}$

Identify the units in the answer. $\frac{4 \text{ miles}}{1 \text{ hour}} \times \frac{}{} = \frac{\text{feet}}{\text{hour}}$

Fill in the conversion rate using equivalent measurements.
1 mile = 5280 ft

$$\frac{4 \text{ miles}}{1 \text{ hour}} \times \frac{5280 \text{ feet}}{1 \text{ mile}} = \frac{21120 \text{ feet}}{1 \text{ hour}}$$

4 miles per hour = 21,120 feet per hour

Make sure the "miles" cancel by having "miles" once in the numerator and once in the denominator.

EXAMPLE 4 Convert 3 feet per minute to feet per hour.

SOLUTION

Write the rate as a fraction. $\qquad \dfrac{3 \text{ feet}}{1 \text{ minute}}$

Identify the units in the answer. $\qquad \dfrac{3 \text{ feet}}{1 \text{ minute}} \times \dfrac{}{} = \dfrac{\text{feet}}{\text{hour}}$

Fill in the conversion rate using equivalent measurements. 1 hour = 60 minutes $\qquad \dfrac{3 \text{ feet}}{1 \cancel{\text{ minute}}} \times \dfrac{60 \cancel{\text{ minutes}}}{1 \text{ hour}} = \dfrac{180 \text{ feet}}{1 \text{ hour}}$

3 feet per minute = 180 feet per hour

EXERCISES

1. Efran can ride his skateboard at a rate of 6 miles per hour. Copy the table and fill in the total miles he skateboards after each hour.

Hours	1	2	3	4	5	6
Total miles traveled						

For Exercises 2–8, show all work necessary to justify your answer.

2. Olivia rode her bike at a rate of 12 miles per hour. She rode at this rate for 2 hours. How far did she ride?

3. Jean walked to her friend's house at a rate of 4.5 miles per hour. She walked for 0.5 hours. How far did she walk?

4. Janette watched a woolly caterpillar crawl across the playground at a rate of 6 inches per minute. It crawled at that rate for 10 minutes. How far across the playground did it crawl?

5. Maricela watched a race car go around a track at a speed of 120 miles per hour. Maricela watched the race car travel 180 miles. How many hours did Maricela watch the race car?

6. A turtle walks 2 feet per minute. How long will it take the turtle to walk 15 feet?

7. Hector ran 15 miles in 3 hours. How far could he run at that speed in 4 hours?

8. Ben drove 25 miles in 0.5 hours. How far could he drive at that speed in 3 hours?

62 Lesson 2.6 ~ Motion Rates

Compare the two rates by finding the unit rate of each. Identify the faster rate.

9. $\dfrac{9 \text{ miles}}{1 \text{ hour}}$ or $\dfrac{16 \text{ miles}}{2 \text{ hours}}$

10. $\dfrac{32 \text{ centimeters}}{2 \text{ minutes}}$ or $\dfrac{120 \text{ centimeters}}{4 \text{ minutes}}$

11. $\dfrac{14 \text{ yards}}{2 \text{ days}}$ or $\dfrac{18 \text{ yards}}{3 \text{ days}}$

12. $\dfrac{5 \text{ kilometers}}{1 \text{ week}}$ or $\dfrac{24 \text{ kilometers}}{4 \text{ weeks}}$

13. Ryan walked at a rate of 5 miles per hour. His sister, Hillary, walked 9 miles in 1.6 hours. Who walked at a faster rate? Explain you you know your answer is correct.

14. Carmen and Gabriella each live 2 miles from the ice rink. Carmen ran to the rink at a rate of 4.2 miles per hour. Gabriella ran to the rink in 0.4 hours. They left their homes at the same time. Who arrived at the ice rink first? Show all work necessary to justify your answer.

15. Two work crews were trying to fix potholes on a highway. The red crew repaired 3 kilometers of highway in 0.5 days. The blue crew repaired 5 kilometers of highway in 1 day. Which crew repaired a longer length of road per day? Use mathematics to justify your answer.

Complete each conversion rate.

16. $\dfrac{1 \text{ kilometer}}{?\ \text{ meters}}$

17. $\dfrac{1 \text{ foot}}{?\ \text{ inches}}$

18. $\dfrac{?\ \text{ minutes}}{1 \text{ hour}}$

19. $\dfrac{1 \text{ meter}}{?\ \text{ centimeters}}$

20. Which of the following rates are conversion rates? Write each answer that applies.

$\dfrac{1 \text{ ft}}{12 \text{ in}}$ $\dfrac{1 \text{ m}}{100 \text{ cm}}$ $\dfrac{2 \text{ yds}}{9 \text{ ft}}$ $\dfrac{1 \text{ minute}}{60 \text{ seconds}}$ $\dfrac{1 \text{ mi}}{5280 \text{ ft}}$ $\dfrac{24 \text{ hours}}{1 \text{ day}}$

Determine which rate should be used to complete each conversion.

21. $\dfrac{8 \text{ miles}}{1 \text{ hour}}$ to $\dfrac{\text{feet}}{\text{hour}}$ A. $\dfrac{1 \text{ mile}}{5280 \text{ feet}}$ or B. $\dfrac{5280 \text{ feet}}{1 \text{ mile}}$

22. $\dfrac{5 \text{ meters}}{1 \text{ minute}}$ to $\dfrac{\text{meters}}{\text{seconds}}$ A. $\dfrac{1 \text{ minute}}{60 \text{ seconds}}$ or B. $\dfrac{60 \text{ seconds}}{1 \text{ minute}}$

23. 2 yards per day to feet per day A. $\dfrac{1 \text{ yard}}{3 \text{ feet}}$ or B. $\dfrac{3 \text{ feet}}{1 \text{ yard}}$

24. A worm travels at a rate of 1 inch per second. Find this rate in inches per minute. Show all work necessary to justify your answer.

Lesson 2.6 ~ Motion Rates

25. A deer runs at a rate of 7 miles per hour.
   a. Convert this rate to feet per hour.
   b. Mark says the deer runs at a rate of 616 feet per minute. Explain how you know Mark is correct.

Write the equivalent rate by converting the rate on the left to the rate on the right.

26. $\frac{2 \text{ miles}}{1 \text{ hour}} = \frac{? \text{ feet}}{1 \text{ hour}}$

27. $\frac{9 \text{ kilometers}}{1 \text{ year}} = \frac{? \text{ meters}}{1 \text{ year}}$

28. $\frac{24 \text{ inches}}{1 \text{ second}} = \frac{? \text{ feet}}{1 \text{ second}}$

29. $\frac{5 \text{ meters}}{1 \text{ minute}} = \frac{? \text{ meters}}{1 \text{ hour}}$

REVIEW

Convert each fraction or mixed number to a decimal. If it is a repeating decimal, use a bar to show which number(s) repeat.

30. $\frac{2}{9}$

31. $3\frac{1}{4}$

32. $1\frac{2}{3}$

33. $\frac{1}{8}$

34. Kirk walked 0.75 miles on the track every morning. Write the distance he walked as a fraction in simplest form.

35. Sierra bought $3.\overline{3}$ pounds of pears at the local farmer's market. Write the weight of her pears as a fraction in simplest form.

36. Jules grew 0.125 inches last month. Write the amount of Jules' growth as a fraction in simplest form.

37. Erin rides her bike 1.6 miles to school everyday. Write the distance she travels as a fraction in simplest form.

38. Locate the 6 and the 18 in the third row of the multiplication table.
   a. Find five different equivalent ratios to 6 : 18 in the chart.
   b. Jake says 2 : 9 is equivalent to 6 : 18. Is he correct? Explain your reasoning.

	1	2	3	4	5	6	7	8	9	10
1	1	2	3	4	5	6	7	8	9	10
2	2	4	6	8	10	12	14	16	18	20
3	3	(6)	9	12	15	(18)	21	24	27	30
4	4	8	12	16	20	24	28	32	36	40
5	5	10	15	20	25	30	35	40	45	50
6	6	12	18	24	30	36	42	48	54	60
7	7	14	21	28	35	42	49	56	63	70
8	8	16	24	32	40	48	56	64	72	80
9	9	18	27	36	45	54	63	72	81	90
10	10	20	30	40	50	60	70	80	90	100

Tic-Tac-Toe ~ Motion Rate

Find a $\frac{1}{4}$ mile track or find a $\frac{1}{4}$ mile length near your home where you can run and walk to answer the questions below.

Step 1: Run $\frac{1}{4}$ mile. Record your time in minutes and seconds. Convert your time to minutes rounded to the nearest hundredth.

Step 2: Find your rate in miles per hour for running $\frac{1}{4}$ mile. Your initial rate will be $\frac{0.25 \text{ miles}}{? \text{ minutes}}$. Convert this rate to miles per hour as a unit rate.

Step 3: Walk $\frac{1}{4}$ mile. Record your time in minutes and seconds. Convert your time to minutes rounded to the nearest hundredth.

Step 4: Find your rate in miles per hour for walking $\frac{1}{4}$ mile. Your initial rate will be $\frac{0.25 \text{ miles}}{? \text{ minutes}}$. Convert this rate to miles per hour as a unit rate.

Step 5: Determine how long it will take you to walk and run each of the lengths below. Assume you keep your calculated rate. Copy and complete the chart.

	1 mile	5 miles	10 miles	26 miles (marathon)
Run time				
Walk time				

Tic-Tac-Toe ~ Fractions and Decimals Game

Create a memory card game using common equivalent fractions and decimals. Common fractions are fractions with denominators of 2, 3, 4, 5, 6, 8 and 10. Use at least four different denominators to create pairs of cards so one card has a fraction and the matching card has the equivalent decimal. The cards should be made on thick paper, such as card stock, construction paper, index cards or poster board. The game must have a minimum of 24 cards.

Lesson 2.6 ~ Motion Rates

REVIEW   BLOCK 2

Vocabulary

equivalent fractions   rate   terminate
motion rates   rate conversion   unit rate
                        repeating decimal

- Convert fractions to decimals and decimals to fractions.
- Convert fractions to repeating decimals.
- Round numbers to a given place value.
- Convert rates to unit rates.
- Solve problems using equivalent rates and unit rates.
- Compare rates using equivalent rates and unit rates.
- Solve problems, compare and convert motion rates.

Lesson 2.1 ~ Fractions and Decimals

Convert each fraction or mixed number to a decimal.

1. $\frac{1}{5}$
2. $\frac{5}{8}$
3. $2\frac{3}{10}$

Determine which fraction is larger by first rewriting the fractions as decimals to compare them.

4. $\frac{3}{8}$ and $\frac{1}{2}$
5. $\frac{3}{12}$ and $\frac{1}{5}$
6. $\frac{4}{5}$ and $\frac{3}{4}$

Convert each decimal to a fraction in simplest form.

7. 0.6
8. 0.5
9. 0.125

10. Petra's mom bought 0.375 pounds of cashews. What is the weight of the cashews as a fraction in simplest form?

Lesson 2.2 ~ Repeating Decimals and Rounding

Convert each fraction to a decimal. Use a bar to show which number(s) repeat.

11. $\frac{2}{3}$
12. $\frac{1}{3}$
13. $\frac{5}{9}$

66   Block 2 ~ Review

Round each number to the nearest tenth.

14. $0.\overline{6}$

15. $0.\overline{45}$

16. $\frac{1}{3}$

17. $\frac{1}{6}$

18. $6.\overline{9}$

19. $\frac{5}{7}$

Round each number to the nearest hundredth.

20. $0.\overline{09}$

21. $\frac{1}{8}$

22. $\frac{2}{3}$

23. Lonnie and Porter were both thinking of decimals written to the hundredths place. Both decimals rounded to the nearest tenth as 5.6. Lonnie's number was the largest possible number that rounded to 5.6 and Porter's was the smallest possible number that rounded to 5.6. What were Lonnie's and Porter's numbers? Explain how you know your answers are correct.

24. Troy ran a race in 12.66 seconds. James ran the same race in $12\frac{2}{3}$ seconds. James said they tied because $12\frac{2}{3} = 12.66$. Is James correct? Explain your reasoning.

Lesson 2.3 ~ Rates and Unit Rates

Find each unit rate.

25. $\frac{450 \text{ miles}}{15 \text{ gallons}}$

26. $\frac{14 \text{ days}}{2 \text{ jobs}}$

27. $\frac{40 \text{ miles}}{4 \text{ hours}}$

28. $\frac{36 \text{ meters}}{9 \text{ minutes}}$

29. $\frac{\$5.50}{11 \text{ pictures}}$

30. $\frac{18 \text{ ounces}}{1.5 \text{ servings}}$

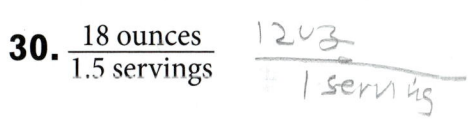

31. Darlene spent $52 for 13 notebooks. Find the price per notebook.

32. Tim bought a box of 36 marbles for $12.00. Find the price per marble. Round your answer to the nearest hundredth.

33. Rebecca used 8 gallons of gas to drive 204 miles. Find how many miles per gallon Rebecca's car traveled. Show all work necessary to justify your answer.

34. Anna skipped at a rate of 2 miles per hour. Her friend, Jenna, skipped at a rate of 0.5 miles in 0.2 hours. Which one skipped at a faster rate? Explain how you know your answer is correct.

35. Lia's car used 5 gallons of gasoline to travel 105 miles. She determined that meant her car had traveled at a unit rate of 21 miles per gallon. Her dad said she traveled at a unit rate of about 0.05 gallons per mile. Explain how both Lia and her dad are correct.

36. Anastasia bought 12 pencils for $6.00. Her work at right shows how much she spent per pencil. Anastasia made a mistake. Explain her mistake and then find the correct amount Anastasia spent per pencil.

$$\begin{array}{r} 2 \\ 6\overline{)12} \\ -12 \\ \hline 0 \end{array}$$

I spent $2.00 per pencil.

Lesson 2.4 ~ Rate Problem Solving

Complete each equivalent rate.

37. $\dfrac{80 \text{ words}}{1 \text{ minute}} = \dfrac{\text{words}}{5 \text{ minutes}}$

38. $\dfrac{12 \text{ kilometers}}{1 \text{ hour}} = \dfrac{\text{kilometers}}{3 \text{ hours}}$

39. $\dfrac{\$2.80}{1 \text{ gallon}} = \dfrac{\$}{10 \text{ gallons}}$

40. $\dfrac{\$5.50}{1 \text{ ticket}} = \dfrac{\$27.50}{\text{tickets}}$

Use equivalent rates to complete each problem. Use mathematics to justify your answer.

41. Holly drove 110 miles in 2 hours. At this rate, how far will she drive in 6 hours?

42. Ari paid $48.33 for 3 concert tickets for herself and two friends. How much was each ticket?

Find each unit rate. Round to the nearest hundredth, if necessary.

43. $\dfrac{\$10.00}{3 \text{ toy cars}}$

44. $\dfrac{25 \text{ millimeters}}{5 \text{ seconds}}$

45. $\dfrac{154 \text{ miles}}{7 \text{ gallons}}$

Use a unit rate to complete each problem. Show all work necessary to justify your answer.

46. Isabella loves eating apples. She eats apples at the rate of 45 apples every 30 days. At this rate, how many apples does Isabella eat in 10 days?

47. Lisa bought pears at a fruit stand. They were on sale for $2.50 for 2 pounds. Lisa bought 9 pounds. How much did Lisa pay for the pears?

Lesson 2.5 ~ Comparing Rates

Use unit rates to determine which of the two rates is smaller.

48. $\dfrac{\$16.00}{5 \text{ toys}}$ or $\dfrac{\$9.75}{3 \text{ toys}}$

49. $\dfrac{11 \text{ miles}}{1 \text{ hour}}$ or $\dfrac{42 \text{ miles}}{4 \text{ hours}}$

50. $\dfrac{10 \text{ calories}}{30 \text{ ounces}}$ or $\dfrac{20 \text{ calories}}{50 \text{ ounces}}$

51. Quinn spent $31.00 for 5 comic books. His friend, Casey, spent $18.00 for 3 comic books. Which person got a better deal? Explain how you know your answer is correct.

52. Janelle paints at a rate of 5 pictures every 8 days. Her teacher paints at a rate of 3 pictures every 6 days. Which person paints more pictures per day? Show all work necessary to justify your answer.

Use equivalent rates to determine the total cost for 12 pounds of tomatoes from two local farms given the price each farm charges.

53. $1.50 per pound of tomatoes

54. $4.00 for 3 pounds of tomatoes

55. Tyler needs to buy 24 pens. He could buy packages of 6 pens for $12.00 or packages of 8 pens for $14.00.
   a. How much will Tyler pay if he chooses to buy packages of 6 pens?
   b. How many packages of the 8 pens would Tyler need to buy?
   c. Which packages should Tyler buy if he wants the cheaper cost? Use mathematics to justify your answer.

Lesson 2.6 ~ Motion Rates

56. Ellen rode her bike at a rate of 15 miles per hour. She rode at this rate for 3 hours. How far did she ride?

57. Stacie walked to school from home at a rate of 3.5 miles per hour. She walked for 0.2 hours. How far is the school from her house?

58. Manuel ran 16 miles in 2 hours. At that rate, how many hours would it take him to run 24 miles? Show all work necessary to justify your answer.

59. One frog hopped at a rate of 3 meters every 15 minutes. A second frog hopped at a rate of 1 meter every 4 minutes. The two frogs entered a frog hopping contest. Which frog won? Explain how you know your answer is correct.

Complete each conversion rate.

60. $\dfrac{1 \text{ meter}}{? \text{ centimeters}}$

61. $\dfrac{1 \text{ yard}}{? \text{ feet}}$

62. $\dfrac{? \text{ seconds}}{1 \text{ minute}}$

Which conversion rate should be used to convert:

63. $\dfrac{45 \text{ miles}}{1 \text{ hour}}$ to $\dfrac{\text{feet}}{\text{hour}}$? **A.** $\dfrac{1 \text{ mile}}{5280 \text{ feet}}$ or **B.** $\dfrac{5280 \text{ feet}}{1 \text{ mile}}$

64. $\dfrac{9 \text{ meters}}{1 \text{ minute}}$ to $\dfrac{\text{meters}}{\text{second}}$? **A.** $\dfrac{1 \text{ minute}}{60 \text{ seconds}}$ or **B.** $\dfrac{60 \text{ seconds}}{1 \text{ minute}}$

65. A dog runs at a rate of 5 miles per hour.
   a. Convert this rate to feet per hour.
   b. Shyanne says the dog runs at a rate of 440 feet per minute. Explain how you know Shyanne is correct.

Write each equivalent rate by converting the rate on the left to the rate on the right.

66. $\dfrac{2 \text{ feet}}{1 \text{ hour}} = \dfrac{\text{inches}}{1 \text{ hour}}$

67. $\dfrac{8 \text{ kilometers}}{1 \text{ minute}} = \dfrac{\text{meters}}{1 \text{ minute}}$

Block 2 ~ Review **69**

CAREER FOCUS

**BRIAN
COACH**

I am a volleyball coach. I do many things to make sure that our team is as successful as they can be. I plan and oversee practice each day. I evaluate my players and decide who is playing the best at what position. I also oversee the assistant coaches on my team.

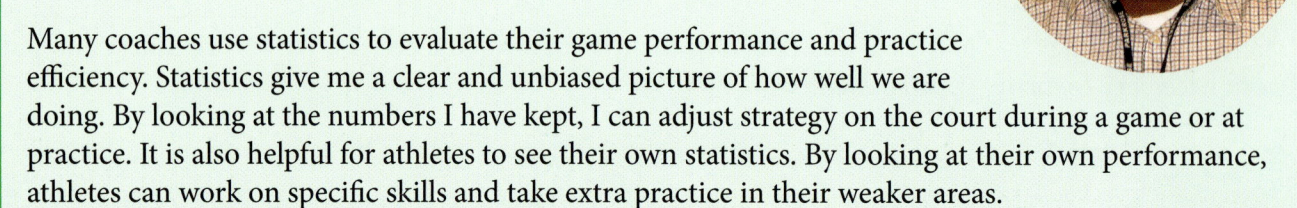

Many coaches use statistics to evaluate their game performance and practice efficiency. Statistics give me a clear and unbiased picture of how well we are doing. By looking at the numbers I have kept, I can adjust strategy on the court during a game or at practice. It is also helpful for athletes to see their own statistics. By looking at their own performance, athletes can work on specific skills and take extra practice in their weaker areas.

In volleyball, players rotate through positions on the court. I use math to analyze rotations and determine the most effective line-up for our team. Two other important components of volleyball are the points my team has scored and the points we have given up. By tracking those numbers throughout the season, I can evaluate our team's trends and help better our odds of competing.

There are no educational requirements for being a coach. Many coaches have received training or college degrees in exercise or fitness management. You can even get a coaching minor at some colleges. Other coaches work their way up through experience.

Salaries for coaches vary greatly depending on the level coached. Coaches of middle or high schools can make up to $6,000 per season. Coaches at the professional level can make millions of dollars per year.

I really enjoy being a coach. It is a daily challenge to come up with a practice and game plan to help my team be as successful as possible. It is also great to see individual athletes work hard and improve. The relationships and experiences I have while coaching make it a very rewarding career.

CORE FOCUS ON RATIOS, RATES & STATISTICS
BLOCK 3 ~ PERCENTS AND PROBABILITY

LESSON 3.1	INTRODUCING PERCENTS	73
	EXPLORE! PERCENTS	
LESSON 3.2	PERCENTS, DECIMALS AND FRACTIONS	77
	EXPLORE! KIERAN'S ROOM	
LESSON 3.3	PERCENT OF A NUMBER	81
LESSON 3.4	PERCENT APPLICATIONS	85
	EXPLORE! AT THE RESTAURANT	
LESSON 3.5	INTRODUCTION TO PROBABILITY	90
LESSON 3.6	EXPERIMENTAL PROBABILITY	94
	EXPLORE! ROLLING A 3	
LESSON 3.7	THEORETICAL PROBABILITY	100
	EXPLORE! SUM OF TWO NUMBER CUBES	
LESSON 3.8	GEOMETRIC PROBABILITY	105
	EXPLORE! WHAT ARE MY CHANCES OF WINNING?	
REVIEW	BLOCK 3 ~ PERCENTS & PROBABILITY	109

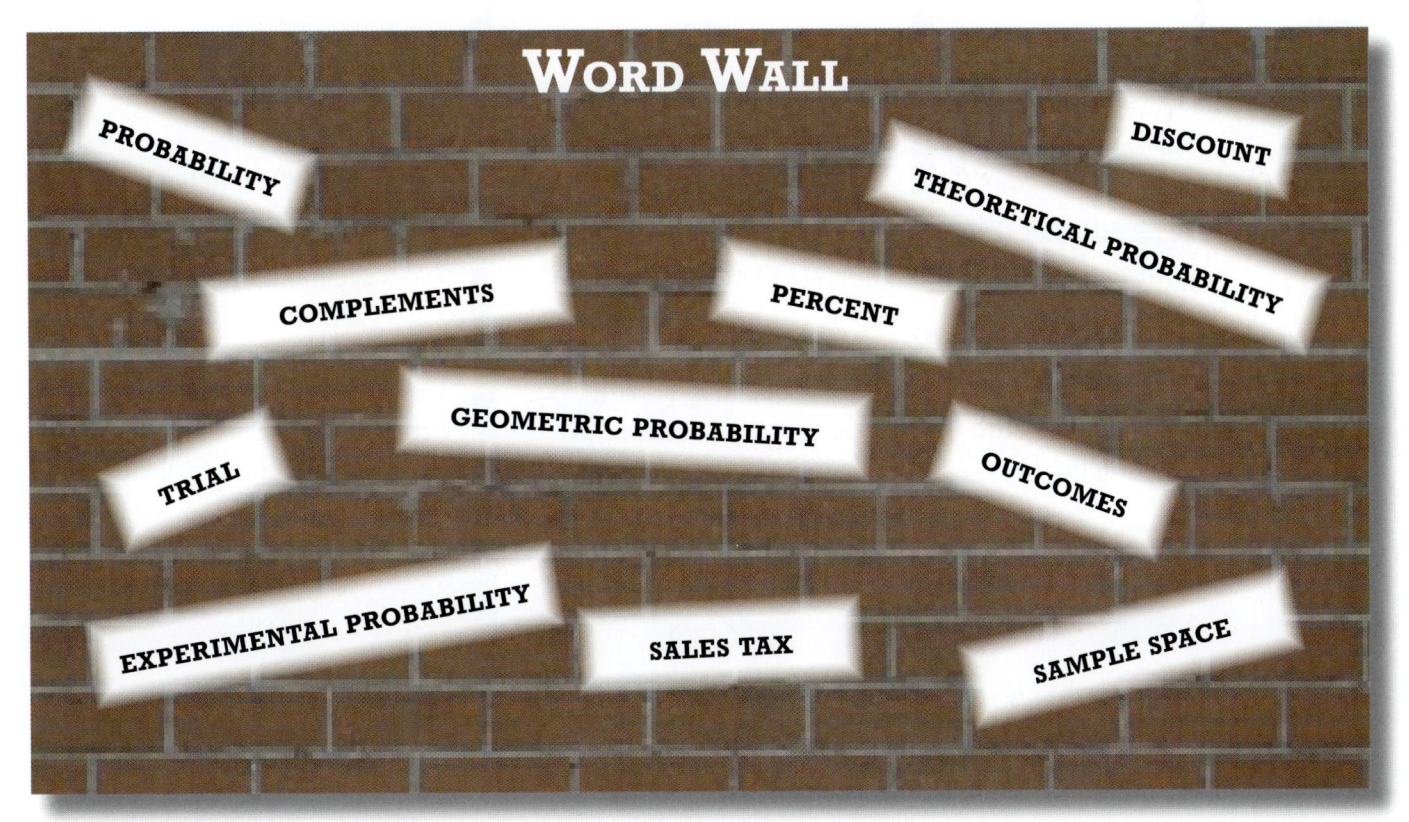

WORD WALL

PROBABILITY

DISCOUNT

THEORETICAL PROBABILITY

COMPLEMENTS

PERCENT

GEOMETRIC PROBABILITY

TRIAL

OUTCOMES

EXPERIMENTAL PROBABILITY

SALES TAX

SAMPLE SPACE

BLOCK 3 ~ PERCENTS AND PROBABILITY
TIC-TAC-TOE

Eye Color

Ask students their eye color. Find the experimental probability for each color.

See page 99 for details.

Carnival Game

Find the geometric probability of winning a carnival game with darts.

See page 108 for details.

Sales Tax

Find the sales tax in different states. Use the information to compare costs of items.

See page 84 for details.

Enough Money?

Explore what happens when an item is on sale and purchased in a state with sales tax.

See page 89 for details.

Menu

Find the amount of tip you should leave at a local restaurant based on the items you order.

See page 89 for details.

Jelly Beans

Find the probability of reaching into a bag and grabbing certain jelly beans.

See page 99 for details.

Rainfall

Find the probability it will rain each month in Seattle and three other cities.

See page 98 for details.

Flap Book

Create a flap book to show other students how to mentally find percents using shortcuts.

See page 84 for details.

Number Cube Differences

Explore the theoretical and experimental probability of rolling two number cubes.

See page 104 for details.

INTRODUCING PERCENTS

LESSON 3.1

◎ Write percents as fractions and decimals.

A **percent** is a ratio that compares a number to 100. When a number is written as a percent, the symbol % is placed after the number. For example, the ratio $\frac{10}{100}$ can also be written 10%.

One way to visualize a percent is to shade squares on a 10 by 10 grid. A 10 by 10 grid has 100 squares so 10% means 10 of the 100 squares are shaded.

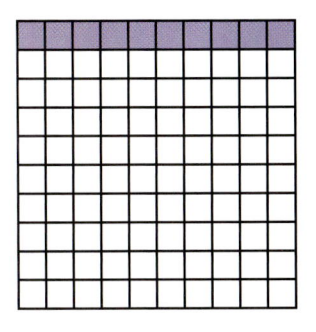

PERCENTS

EXPLORE!

Step 1: For each shaded grid, write:
- The ratio of the shaded squares to 100 (a fraction).
- The percent of squares shaded as a number with the % sign.

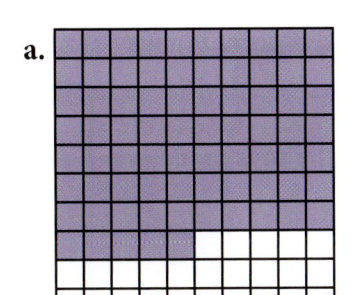

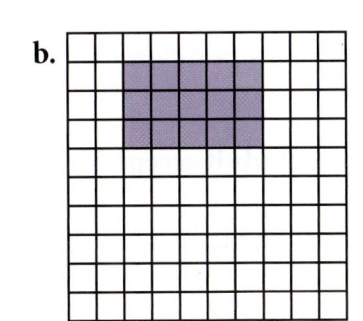

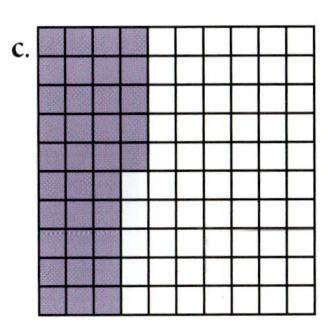

a.   b.   c.

Step 2: How many squares would be shaded on a 10 by 10 grid for each problem?

  a. 1%   b. 25%   c. 50%   d. 100%   e. 0%

Step 3: Kim bought 100 balloons for her birthday party. She used 86 of them. What percent of the balloons did she use?

Step 4: B.J. used 60 envelopes out of 100.
  a. What percent of envelopes did he use?
  b. What percent of envelopes were left over?

Step 5: Create a 10 by 10 grid. Shade in $\frac{3}{10}$ of the squares.

Step 6: What percent of the squares are shaded in the grid from **Step 5**?

Step 7: Write $\frac{3}{10}$ as a decimal. What do you notice about the decimal and the percent from **Step 6**?

Lesson 3.1 ~ Introducing Percents   73

Write each percent as a decimal.

14. 10% **15.** 40% **16.** 33%

17. 82% **18.** 5% **19.** 50%

20. 400% **21.** 0.5% **22.** 125%

23. Trace the two circles.
   a. Shade 100% of one circle.
   b. Shade 100% of the second circle.
   c. Together, you shaded _____ circles.
   d. One shaded circle represents 100%. What percent is represented by the shaded regions in the circles you have drawn?

24. Margaret looked at the juice box her mom packed in her lunch. The box said it was 10% real fruit juice. Margaret said that meant 90% of the drink was not real fruit juice. Explain why Margaret is correct.

25. Fred found a sale on a new jacket he wanted. It was on sale for 15% off the original price. What percent of the original price was Fred going to pay for the jacket?

26. Olga added on to her home until it was 300% as big as when she bought it.
   a. Write this percent as a fraction.
   b. Write this percent as a decimal.
   c. Explain what it means if Olga's home is 300% as big as when she bought it.

27. Leonard needed a part to fix his washing machine. He looked for the part online. He found that only 0.7% of the stores had the part in stock.
   a. Write this percent as a fraction without decimals.
   b. Write this percent as a decimal.
   c. Explain what it means for 0.7% of the stores to have the part in stock.

28. One serving of cereal contains 25% of the daily recommended amount of calcium. How many servings of cereal would someone need to eat to get 100% of the daily recommended amount of calcium?

29. Draw a 10 by 10 grid on graph paper.
   a. Design a room on the grid with the following: the bed is 30% of the room, the dresser is 24% of the room and the desk is 20% of the room.
   b. What percent of the room is left over?

REVIEW

Complete each conversion.

30. 3 miles = _____ feet

31. 4 tons = _____ pounds

32. 35 days = _____ weeks

33. 250 meters = _____ kilometers

76 Lesson 3.1 ~ Introducing Percents

PERCENTS, DECIMALS AND FRACTIONS

LESSON 3.2

Write fractions and decimals as percents.

In **Lesson 3.1** you learned to write percents as fractions or decimals. You will learn to convert a fraction or decimal to a percent in this lesson.

EXPLORE! KIERAN'S ROOM

Kieran used a piece of 4 by 5 grid paper to sketch the floor plan of his room. He colored the location of his bed with purple and the location of his dresser with blue. The space his desk occupies is colored green.

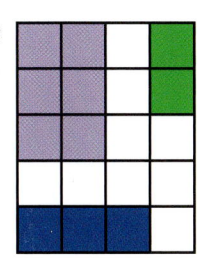

Step 1: Write the ratio of the parts shaded for each object to the total space as a fraction in simplest form.
  a. Bed **b.** Dresser **c.** Desk

Step 2: Convert each fraction in **Step 1** to a decimal.

Step 3: A decimal can be converted to a percent by multiplying by 100. Multiply each decimal in **Step 2** by 100 to determine what percentage of the room is taken up by each object.

Step 4: Kieran's sister, Tamika, gave Kieran an incomplete chart describing her room. Copy and complete the chart for Tamika's room.

	Bed	Dresser	Desk
Fraction			$\frac{1}{20}$
Decimal		0.1	
Percent	40%		

Step 5: Draw a new 4 by 5 grid for Tamika's room. Shade in the correct number of squares for each piece of furniture. Will your grid look exactly the same as other classmates' grids? Explain your reasoning.

There are two ways to write a decimal as a percent. One method rewrites the decimal as a ratio comparing a number to 100. The second method multiplies the decimal by 100. Both are shown in the following examples.

Lesson 3.2 ~ Percents, Decimals and Fractions 77

EXAMPLE 1   Write each decimal as a percent.
a. 0.2   b. 1.63

SOLUTIONS

Method 1 Rewrite as a fraction $\frac{?}{100}$	Method 2 Multiply by 100
a. 0.2 Write the decimal as a fraction. $\frac{2}{10}$ Write an equivalent fraction $\frac{2 \times 10}{10 \times 10} = \frac{20}{100}$ with 100 as the denominator. Write the percent. $\frac{20}{100} = 20\%$	**a.** 0.2 Multiply by 100. $0.2 \times 100 = 20$ Write with the percent symbol. 20%
b. 1.63 Write the decimal as a fraction. $\frac{163}{100}$ Write the percent. $\frac{163}{100} = 163\%$	**b.** 1.63 Multiply by 100. $1.63 \times 100 = 163$ Write with the percent symbol. 163%

EXAMPLE 2   Write each fraction as a percent.

a. $\frac{2}{5}$ b. $\frac{1}{3}$

SOLUTIONS

Method 1 Rewrite as a fraction $\frac{?}{100}$	Method 2 Multiply the decimal by 100
a. Write an equivalent fraction $\frac{2 \times 20}{5 \times 20} = \frac{40}{100}$ with 100 as the denominator. Write the percent. $\frac{40}{100} = 40\%$	**a.** Write the fraction as $\frac{2}{5} = 2 \div 5 = 0.4$ a decimal. Multiply by 100. $0.4 \times 100 = 40$ Write with the percent symbol. 40%
b. 3 does not divide evenly $\frac{1 \times ?}{3 \times ?} = \frac{?}{100}$ into 100 so use Method 2.	**b.** Write the fraction as $\frac{1}{3} = 1 \div 3 = 0.\overline{3}$ a decimal. Multiply by 100. $0.\overline{3} \times 100 = 33.\overline{3}$ Write with the percent symbol. $33.\overline{3}\%$

Lesson 3.2 ~ Percents, Decimals and Fractions

EXERCISES

Write the shaded part of each shape as a fraction in simplest form, a decimal and a percent.

1.
2.
3.
4.
5.
6.

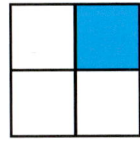

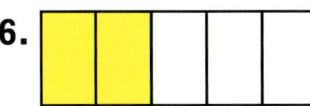

Write each decimal as a percent.

7. 0.04

8. 0.6

9. 0.565

10. 0.1

11. 2.25

12. 4.5

Write each fraction as a percent.

13. $\frac{1}{2}$

14. $\frac{3}{10}$

15. $\frac{2}{3}$

16. $\frac{3}{8}$

17. $1\frac{1}{4}$

18. $4\frac{1}{2}$

19. Copy the grid at right on your paper.
    a. Shade $\frac{1}{4}$ of the squares.
    b. What percent of the squares are shaded?
    c. What percent of the squares are not shaded?

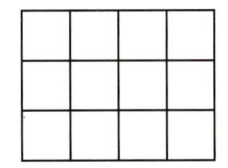

20. Copy the grid at right on your paper.
    a. Shade $0.\overline{6}$ of the squares.
    b. What percent of the squares are shaded?
    c. What percent of the squares are not shaded?

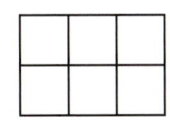

21. Write each value from the list below that is equivalent to seventy-five hundredths.
    0.75 $\frac{3}{4}$ 75%   25%   7.5 $\frac{75}{100}$

22. Write each value from the list below that is equivalent to one-third.
    0.3 $\frac{1}{3}$ $33\frac{1}{3}$%   30%   0.03 $0.\overline{3}$

23. Mona's work for converting 0.4 to a percent is below. Her work is correct. Show another way to find the answer.

    $$0.4 \times 100 = 40$$
    $$0.4 = 40\%$$

Lesson 3.2 ~ Percents, Decimals and Fractions **79**

Order each group of numbers from least to greatest.

24. 25%, 0.24, $\frac{1}{5}$

25. $\frac{2}{3}$, 60%, 0.7

26. 1.2, 130%, $1\frac{1}{4}$

27. At a movie premier in Hollywood, $\frac{9}{20}$ of the people who attended liked the film.
   a. What percent of the people who attended the movie premier liked the film?
   b. What percent of the people who attended the movie premier did not like the film?

28. Five of the nine lighthouses along the coast of Oregon are in use. The others are designated historic monuments.
   a. Write the ratio of the number of lighthouses used to the total number along the coast as a fraction.
   b. Write the fraction in **part a** as a decimal.
   c. What percent of the lighthouses along the Oregon coast are currently in use?

29. Dillon asked 50 students at school what type of music they preferred. He found that $\frac{2}{5}$ of the students preferred rock music. What percent of the students preferred rock music? Show all work necessary to justify your answer.

30. Copy and complete the table.

Fraction	$\frac{1}{5}$		$\frac{1}{3}$			$\frac{3}{5}$		$\frac{3}{4}$		
Decimal		0.25			0.5				0.8	
Percent				40%			$66\frac{2}{3}$%			100%

REVIEW

Write each percent as a fraction.

31. 25%

32. 30%

33. 45%

34. Zarina bought a sweater for 60% off the original price. What fraction of the original price did she pay?

Write each percent as a decimal.

35. 33%

36. 0.7%

37. 280%

38. Damon had a jar of pennies which was 45% full. Write this percent as a decimal.

39. The ratio of red to blue marbles is the same in three jars. Each ratio is shown in the table. Find the missing value. Explain your reasoning.

red	4	8	?
blue	5	10	30

PERCENT OF A NUMBER

LESSON 3.3

Find the percent of a number.

A jacket is on sale for 25% off the original price of $40. You need to figure out what 25% of 40 is to find out how much money will be taken off the original price. In this lesson you will find percents of numbers by rewriting percents as fractions or decimals.

EXAMPLE 1 Solve each percent problem.
   a. 25% of 40 is _____   b. _____ is 60% of 30

"of" means multiply (×) and "is" means equals (=).

SOLUTIONS

a. 25% of 40 is _____
   25% × 40 = _____

   Use a fraction. $\frac{1}{4} \times 40 = \frac{1}{4} \times \frac{40}{1} = \frac{40}{4} = 10$
   OR
   Use a decimal. $0.25 \times 40 = 10$

   25% of 40 is 10.

b. _____ is 60% of 30
   _____ = 60% × 30

   Use a fraction. $\frac{3}{5} \times 30 = \frac{3}{5} \times \frac{30}{1} = \frac{90}{5} = 18$
   OR
   Use a decimal. $0.60 \times 30 = 18$

   18 is 60% of 30.

Percents are very useful when working with money. You use percents to find the value of tips, taxes and sale prices.

EXAMPLE 2 Find the value of each expression.
   a. 40% of $1.00   b. 75% of $200

SOLUTIONS

a. Convert the percent to a fraction or decimal. $40\% = 0.4 = \frac{40}{100} = \frac{2}{5}$

   Find the value of the expression using a decimal or fraction.

   $0.4 \times 1.00 = \$0.40$
   OR
   $\frac{2}{5} \times \frac{1.00}{1} = \$0.40$

   40% of $1.00 is $0.40.

Lesson 3.3 ~ Percent of a Number   81

EXAMPLE 2
SOLUTIONS
(CONTINUED)

b. Convert the percent to a fraction or decimal.

$75\% = 0.75 = \frac{75}{100} = \frac{3}{4}$

Find the value of the expression using a decimal or fraction.

$0.75 \times 200 = \$150$

OR

$\frac{3}{4} \times \frac{200}{1} = \$150$

75% of $200 is $150.

EXAMPLE 3

The price of gasoline this week is 105% of the price last week. Find the price of gasoline this week if gasoline was $4.00 per gallon last week.

SOLUTION

Write the problem.   105% of $4.00

Write the percent as a decimal. $105\% = 1.05$

Write an expression and simplify. $1.05 \times 4.00 = 4.20$

The price of gasoline is $4.20 per gallon this week.

EXAMPLE 4

Draw a line segment that is 75% as long as the line segment below.

SOLUTION

Measure the line segment using centimeters.   6 cm

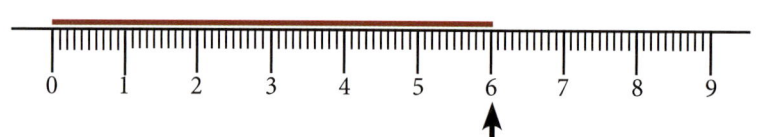

Convert the percent to a decimal. $75\% = 0.75$

Multiply the length by 0.75. $6 \times 0.75 = 4.5$

Draw a line segment which is 4.5 cm in length.

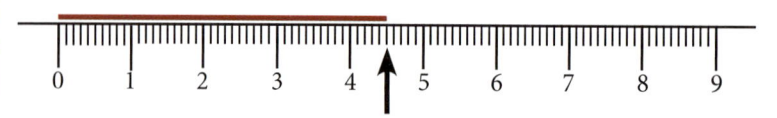

EXERCISES

Solve each percent problem.

1. 10% of 70 is _____

2. 25% of 60 is _____

3. 15% of 20 is _____

4. _____ is 5% of 90

5. _____ is 120% of 25

6. 30% of 66 is _____

7. 22% of 200 is _____

8. _____ is 90% of 50

9. _____ is 56% of 20

82   *Lesson 3.3 ~ Percent of a Number*

10. Find 42% of $1.00.

11. Find 15% of $3.00.

12. What percent of $1.00 is $0.30?

13. Find 75% of 100 meters.

14. Sandra recorded the lengths of her frog's three best hops: 25 *cm*, 30 *cm* and 15 *cm*. The total distance her frog hopped is what percent of a meter?

15. What percent of 1 meter is 56 centimeters? Explain how you know your answer is correct.

Complete each table.

16.
Number	10	20	30	60	80
10% of Number	1	2	3		

17.
Number	20	24	32	40	80
25% of Number	5	6	8		

18.
Number	10	16	30	50	80
50% of Number	5	8	15		

19.
Number	9	21	30	60	90
$33\frac{1}{3}$% of Number	3	7	10		

20. There are shortcuts for finding common percentages like 10%, 25%, 50% and $33\frac{1}{3}$% as in **Exercises 16-19**. Describe the shortcut for each.

Draw a line segment that fits each description. Record the length of each new line segment.

21. 50% of the line segment

22. 25% of the line segment

23. 200% of the line segment

24. 75% of the line segment

25. Sara says 30% of 60 is 20 because 2 × 3 = 6. Is she correct? Explain your reasoning.

REVIEW

Write each decimal or fraction as a percent.

26. 0.4

27. 1.35

28. 0.001

29. $\frac{62}{100}$

30. $\frac{11}{20}$

31. $\frac{1}{4}$

32. Use the circle.
   a. What fraction of the circle is shaded?
   b. What fraction of the circle is not shaded?
   c. What is the sum of the fractions in **a and b**?
   d. What percent of the circle is shaded?
   e. What percent of the circle is not shaded?
   f. What is the sum of the percents in **d and e**?

Lesson 3.3 ~ Percent of a Number

Tic-Tac-Toe ~ Flap Book

There are shortcuts to find some percents.

Example: Find 10% of 32.

$0.1 \times 32 = 3.2$ which is the same as $\frac{1}{10} \times 32 = \frac{32}{10} = 32 \div 10 = 3.2$

In other words, to find 10% of 32, you can find $32 \div 10$.

Step 1: Find 10% of each number without using a calculator.

    a. 100   **b.** 50   **c.** 30   **d.** 12   **e.** 71

There are similar shortcuts for 20%, 25% and 50%. Use the example above to find each shortcut.

Step 2: What number do you need to divide by to find 20% of a number?

Step 3: What number do you need to divide by to find 25% of a number?

Step 4: What number do you need to divide by to find 50% of a number?

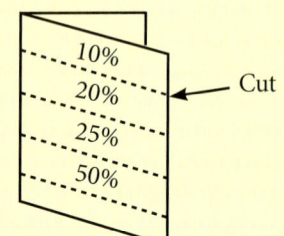

Step 5: Create a flap book that explains each of the shortcuts for finding 10%, 20%, 25% and 50%. This book may be used to tutor another student. Include examples in your book. Find one other percent with a shortcut like **Step 1**. Include this in your book.

Tic-Tac-Toe ~ Sales Tax

Most states have a sales tax which is applied to non-grocery items. If a state has an 8% sales tax, you pay the cost of the item and an additional 8% of the item's cost.

Example: An item costs $100. An 8% sales tax is added to the item cost. You pay $108 total.

Step 1: Research and list the states in the United States that do not charge sales tax.

Step 2: Find the two states which have the highest sales tax.

Step 3: Find the two states which have the lowest sales tax greater than zero.

Step 4: Suppose you want to buy each of the items listed below. Write each item's total cost (including sales tax) if you purchase it in (1) your state, (2) the state with the highest sales tax and (3) the state with the lowest sales tax.

- Laptop computer for $800
- Used car for $6,000
- Television for $1,500

Organize all of your information in a chart.

PERCENT APPLICATIONS

LESSON 3.4

Solve problems involving discounts, tips and tax.

You use percents in many real-world situations. Items in stores may have sale prices based on a certain percent off the original price. The percent helps you calculate the **discount**. A discount is the amount of money subtracted from the original price to give you the sale price.

EXAMPLE 1

Anna ordered a new vacuum cleaner with an original price of $98. The vacuum cleaner was on sale for 20% off.
a. What was the value of the discount?
b. What was the sale price of the vacuum cleaner?

SOLUTIONS

a. Write the problem. 20% of $98 is what?

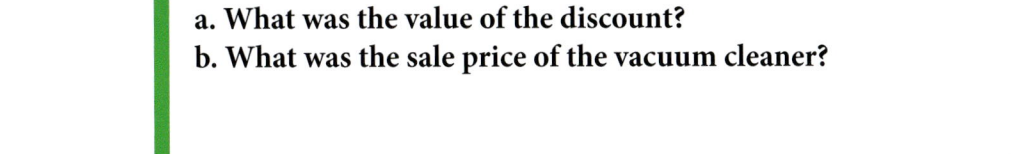

Convert words to a math expression and simplify. $0.20 \times 98 = 19.6$

The value of the discount was $19.60.

b. Anna saved $19.60 when she ordered the vacuum cleaner. Find the amount she paid by subtracting the amount she saved from the total.

```
  98.00
- 19.60
  78.40
```

The sale price of the vacuum cleaner was $78.40.

EXAMPLE 2

Bryan went to the mall to buy a new video game. When he got to the store he found there was a sale. The game was originally $29.99. It had a sticker which read, "10% off the original price!" What was the value of the discount?

SOLUTION

Write the problem. 10% of 29.99 is what?

Convert the words to a math problem. $0.10 \times 29.99 = 2.999$

Round to the nearest cent. $\$2.999 \approx \$3.00$

The value of the discount was $3.00.

Lesson 3.4 ~ Percent Applications 85

EXPLORE!   **AT THE RESTAURANT**

Nina and Cameron went to Billy's BBQ for lunch. The menu items are shown below.

Billy's BBQ
Ribs	$8.00
Chicken	$6.50
Beef Brisket	$9.00
Baked Beans	$2.00
Coleslaw	$1.50
Roll	$1.00
Soda	$1.50

Step 1: Nina orders chicken, baked beans and a soda. What would be the cost of her meal?

Step 2: Cameron orders ribs, coleslaw, a roll and a soda. What would be the cost of his meal?

Step 3: A tip for a server is based on the amount of the bill. It is common to leave 15% of the total bill as a tip. Find the amount Nina should leave as a tip following the 15% rule.

Step 4: Find the amount Cameron should leave as a 15% tip.

Step 5: Cameron pays for both meals. Billy was an excellent server so Cameron leaves a 20% tip. How much did Cameron leave as a tip?

Step 6: What was the total cost for both meals and a 20% tip?

Many states have **sales tax**. It is an amount added to the cost of an item bought in a store. In 2012, Oregon, Delaware, Montana, New Hampshire and Alaska did not have state sales taxes. However, when people from these states visit states with a sales tax, they usually need to pay the sales tax in that state.

EXAMPLE 3 Jose traveled to Florida to visit family. While in Florida he bought a shirt priced at $25.00. The sales tax in Florida was 6%.
a. How much sales tax did he pay for the shirt?
b. What was the total amount he paid for the shirt?

SOLUTIONS

a. Write the problem.   6% of $25.00 is what?

Solve the percent problem.   0.06 × 25.00 = 1.5

Jose will pay an additional $1.50 in sales tax.

b. Add the tax to the original price.   25.00
                                                              + 1.50
                                                              26.50

The total amount for the shirt, including tax, was $26.50.

Lesson 3.4 ~ Percent Applications

EXERCISES

1. Ms. Pearson had a flower garden. In the spring she planted 100 seeds. Of the seeds planted, 35% were pansies. How many pansies did Ms. Pearson plant?

2. Nate collects baseball cards. He has a total of 130 cards. Thirty percent of his cards are players from his favorite team. How many baseball cards does Nate have from his favorite team?

3. Tabitha planted an evergreen tree in her front yard. It was 12 feet tall. The next year it was 20% taller. How much taller, in feet, was the evergreen tree in Tabitha's yard?

4. Mr. Lloyd had 32 students in his 6th grade class. Of those students, 75% were boys. How many boys were in Mr. Lloyd's 6th grade class?

5. A music store in New York was going out of business. The sign read, "Everything must go! 60% off every item!" Kyron bought a CD there that was originally $14.99.
    a. How much was the discount?
    b. How much did Kyron pay for the CD?

6. Doreen went to the store to buy a pair of jeans. The jeans she liked best were originally $40.00. They were on sale for 30% off the original price. What was the sale price for the jeans? Show all work necessary to justify your answer.

7. Carter went to the store to buy a new keyboard. He found last year's model on sale for 15% off the original price, which was $199.00. How much was the discount on the keyboard? Use mathematics to justify your answer.

8. Kari went to the mall and found a store having a clearance sale. Everything in the store was 40% off. She purchased a pair of pants that were originally $35.00 and a jacket that was originally $40.00. What was the total cost of her purchase? Show all work necessary to justify your answer.

Below are several restaurant bills. Each customer left a 15% tip. Find the amount of money left for each tip. Round to the nearest hundredth, if necessary.

9. Total bill: $15.00

10. Total bill: $65.00

11. Total bill: $44.95

12. Mikayla took her friends to lunch. The bill came to $32. She wanted to leave a 15% tip and needed to determine the total cost of the lunch, including the tip. Her work is below. Unfortunately, Mikayla made a mistake. Explain the mistake Mikayla made and then find the correct total for the bill.

    Tip: 32 × 0.15 = 4.80

    Total Lunch Bill: $32.00 − $4.80 = $27.20

13. The Jackson family went to dinner with the Welton family. The total bill for the meal was $88. The families decided to leave a 20% tip and then split the cost of the meal and tip equally. How much did each family spend? Show all work necessary to justify your answer.

14. Michelle visited her sister in California. While there, she bought $42.00 worth of party supplies for her nephew's birthday. The city where she bought the supplies had an 8% sales tax. Find the amount she spent in sales tax on the birthday supplies.

15. Carla went to Idaho and found a car for $24,000. If she buys it in Idaho she will be charged an additional 6% sales tax on the price of the car.
   a. How much sales tax will Carla pay on the car in Idaho?
   b. What is the total price, including tax, for the car?

16. Emie went to Washington DC and stayed with friends in Virginia. While there she bought a souvenir book for $15.00. The sales tax on the book was 5%. Find the total cost of the book, including tax. Show all work necessary to justify your answer.

17. A $20 DVD is discounted 50%. It was then discounted an additional 25%. Is this the same as a 75% discount? Explain your reasoning.

REVIEW

Write each percent as a fraction and a decimal.

18. 20% **19.** 10% **20.** 50%

21. 110% **22.** 0.1% **23.** 82%

24. Convert 3 kilometers per hour to meters per hour.

25. Convert 2 weeks per job to days per job.

26. Chet could buy a package of 8 pencils for $0.72 or a package of 12 pencils for $0.96. Which package is a better deal? Explain your reasoning.

27. Eliza biked 15 miles in 1.5 hours. Yan biked 21 miles in 2 hours. Who biked at a faster rate? Use mathematics to justify your answer.

28. The points in the table have equivalent ratios. Plot the points and graph to find the missing value.

x	6	9	15	21
y	20	30	?	70

Tic-Tac-Toe ~ Enough Money?

Tara needs a new purse. She has $30. She found a purse she wants for $32. It is on sale for 20% off the original price. She must also pay a 5% sales tax. Does she have enough money to buy the purse?

Example:
Step 1: Find the discount on the purse → 0.2 × 32 = 6.4 $ 6.40
Step 2: Find the cost of the purse after the discount is applied → 32 − 6.4 = 25.6 $25.60
Step 3: Find the cost of the 5% sales tax → 0.05 × 25.60 = 1.28 $ 1.28
Step 4: Find the cost of the purse with the sales tax → 25.60 + 1.28 = 26.88 $26.88

The purse will cost $26.88 which means Tara has enough money to buy it.

Use the steps above to find the cost for each item below when purchased in a state with an 8% sales tax.

1. An $800 computer on sale for 10% off the original price.
2. A $300 table on sale for 25% off the original price.
3. A $42 jacket on sale for 30% off the original price.
4. A DVD was on sale for 50% off the original price. It was later discounted an additional 20% off the sale price. Is this the same as a 70% discount? Explain using complete sentences. Include the example of a DVD with an original price of $24.

Tic-Tac-Toe ~ Menu

Find a take-out menu for a local restaurant. Use a menu from the internet or an actual menu from a restaurant. If you use an actual menu, ask the owner for permission to use it on this activity. If using a paper copy of a menu, attach it to your work.

Complete these steps for **Questions 1, 2, 4, 5 and 7**.
  Step 1: Identify the items on the menu and each price.
  Step 2: Find the total price for the items on the menu.
  Step 3: Find the amount of a 15% tip.
  Step 4: Find the total cost for the meal, including the tip.

1. A beverage, main dish and dessert.
2. Water with the same main dish and dessert as in #1.
3. Why do you think a waiter offers you a drink when you first sit down?
4. A beverage, an appetizer and a main dish.
5. The same beverage and main dish as in #4, but no appetizer.
6. Why do you think a waiter offers you an appetizer when you first sit down?
7. You have $20 to spend for lunch, including the tip. Which items can you buy? Identify the items and the total cost, including a tip, for your meal.

INTRODUCTION TO PROBABILITY

LESSON 3.5

Recognize probabilities that are unlikely and likely.
Express probabilities as fractions, decimals and percents.

Bonnie and Juan are preparing to hike along the Appalachian Trail. They are trying to decide whether to hike on Friday or Saturday. Bonnie checked the weather forecast online. The weather report says there is a 90% chance of rain Friday and a 10% chance of rain Saturday. Based on the weather report, on which day is it more likely to rain?

Probability measures how likely it is something will occur. The probability that something will occur can be written as a fraction, decimal or a percent. When written as a percent, the probability something will occur is 0% to 100%. When written as a fraction or a decimal, the probability something will occur is 0 to 1.

The probability that it will rain on Bonnie and Juan's hike was given as a percent. It can also be represented by a fraction or a decimal.

Probability of rain on Friday 90% = $\frac{9}{10}$ = 0.9

Probability of rain on Saturday 10% = $\frac{1}{10}$ = 0.1

The more likely it is something will occur, the closer its probability is to 1. The less likely it is something will occur, the closer its probability is to 0.

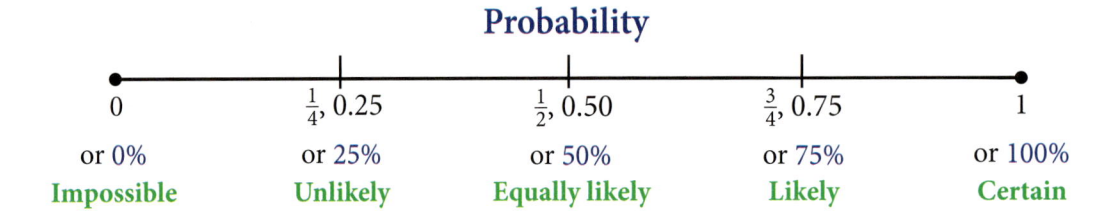

Probability

0	$\frac{1}{4}$, 0.25	$\frac{1}{2}$, 0.50	$\frac{3}{4}$, 0.75	1
or 0%	or 25%	or 50%	or 75%	or 100%
Impossible	Unlikely	Equally likely	Likely	Certain

EXAMPLE 1 Determine whether each event is *impossible, unlikely, equally likely, likely* or *certain*.
a. A card thrown on the floor lands face up.
b. You roll a 2 on a regular number cube.
c. You are older now than when you were born.

SOLUTIONS a. Equally likely. It can land face up or face down – one is not more likely than the other.

b. Unlikely. There are 6 numbers on a number cube and 2 is only one of the numbers.

c. Certain. You are older now than when you were born.

EXAMPLE 2 Dexter's teacher said there was an 80% chance of getting math homework this weekend. Write this probability as a simplified fraction and a decimal.

SOLUTION As a fraction, 80% = $\frac{80}{100}$ which simplifies to $\frac{4}{5}$.

As a decimal, 80% = $\frac{80}{100}$ = 80 ÷ 100 = 0.8.

An 80% chance of math homework is the same as a 0.8 chance of homework or a $\frac{4}{5}$ chance of homework.

EXAMPLE 3 Felipe asked 50 students which type of movie they preferred. Forty percent of the students chose action movies, $\frac{1}{10}$ chose drama and 0.5 chose comedy. How many of the students chose each type of movie?

SOLUTION
Action movies: Find 40% of 50. 0.4 × 50 = 20 people

Drama movies: Find $\frac{1}{10}$ of 50. $\frac{1}{10} \times \frac{50}{1} = \frac{50}{10}$ = 5 people

Comedy movies: Find 0.5 of 50. 0.5 × 50 = 25 people

☑ 20 + 5 + 25 = 50. All 50 people are included.

EXERCISES

Determine whether each event is *impossible, unlikely, equally likely, likely* or *certain*.

1. School will be in session during the month of October.

2. You will roll a 10 on a regular six-sided number cube.

3. One of two equally matched people will win a game of Tug-of-War.

4. It will snow on Mt. Rainier today if there is an 80% chance of snow.

5. It will rain on the Texas plains today if there is a 0.5 chance of rain.

6. Sally will choose a cashew from a bag of mixed nuts that has a ratio of cashews to all nuts of $\frac{10}{21}$.

Lesson 3.5 ~ Introduction to Probability

Use the spinner at right to determine whether each event is *impossible, unlikely, equally likely, likely* or *certain*. Use mathematics to justify your answer.

7. You land on a blue section on the spinner.

8. You land on a yellow section on the spinner.

9. You land on the blue, red or yellow section on the spinner.

10. You land on a green section on the spinner.

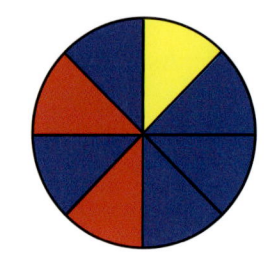

11. Gracie planted flowers in her garden. There is a 25% chance a flower that sprouts will be yellow. Write this probability as a simplified fraction and as a decimal.

12. Describe something that has a probability of 0. Explain why it has a probability of 0.

13. Describe something that has a probability of 1. Explain why it has a probability of 1.

14. Describe something that is unlikely to occur. Explain why it is unlikely to occur.

15. Adila was eating jelly beans. Suppose there is a 70% chance the next jelly bean she picks will be red.
    a. Write this probability as a simplified fraction and as a decimal.
    b. Explain what the bag of jelly beans must look like based on this probability.

16. Matthew wants to play golf on Saturday. The weatherman said there is a 0.35 chance it will rain. Write this probability as a percent and as a simplified fraction.

17. Wade was playing a game. He had a 0.9 chance of choosing the winning card.
    a. Write this probability as a percent and as a simplified fraction.
    b. Do you think he will win the game? Explain your reasoning.

18. Julianne planted red, green, yellow and orange peppers in her garden. There is a $\frac{2}{5}$ chance that the first ripe pepper will be red. Write this probability as a decimal and as a percent.

19. There is a $\frac{2}{3}$ chance the construction on Route 66 will be finished today.
    a. Write this probability as a decimal and as a percent.
    b. Are you confident the construction will be finished today? Explain your reasoning.

20. Reid and Ian each said the probability that a spinner would land on yellow was unlikely. Reid said it was because the probability was $\frac{1}{4}$, but Ian said it was because the probability was 0.25. Explain why Reid and Ian's reasoning is the same.

21. Write each value from the list below that describes a probability which is unlikely.

0.1 $\frac{1}{8}$ 50% $\frac{3}{10}$ 0.4 $\frac{5}{8}$ 0.6   15%

22. Greg randomly polled 100 students to determine their favorite lunch foods from the cafeteria. His results are shown in the table below.

Foods	Hamburgers	Tacos	Spaghetti	Sandwiches	Stir-Fry
Probability each food was chosen	$\frac{1}{4}$	0.15	$\frac{1}{10}$	40%	0.1

a. Write each probability as a fraction.
b. Write each probability as a decimal.
c. Write each probability as a percent.
d. Based on the probabilities, which food(s) did students most prefer?
e. Based on the probabilities, which food(s) did students least prefer?

23. Dave surveyed 40 students to see which flavor of ice cream they preferred. Thirty percent chose strawberry, 0.2 chose chocolate and $\frac{1}{2}$ chose vanilla. List the flavors from least popular to most popular. Use mathematics to justify your answer.

24. Use the spinner to the right.
a. What is the probability you will land on a blue section?
b. What is the probability you will land on a red section?
c. What is the probability you will land on a yellow section?

Three different probabilities are given in each exercise. Order them from least to greatest.

25. 30%, $\frac{1}{3}$, 0.35

26. 0.4, 45%, $\frac{3}{8}$

27. $\frac{4}{5}$, 75%, 0.7

REVIEW

Solve each percent problem.

28. 50% of 20 is _____

29. _____ is 30% of 120

30. 10% of 42 is _____

31. 25% of 50 is _____

32. 200% of 34 is _____

33. 0.1% of 100 is _____

34. Max went to the store to buy a new jacket. He found a jacket he liked with an original price of $80.00. It was on sale for 20% off the original price. How much did Max pay for the jacket? Show all work necessary to justify your answer.

35. Justin and Jen went out for lunch at their favorite restaurant. The bill was $28.95. They left a 15% tip. Find the tip amount they left for the server. Round your answer to the nearest hundredth.

EXPERIMENTAL PROBABILITY

LESSON 3.6

Find and interpret the experimental probability of an event.

Karen wanted to know how many people were going to put tomatoes on their tacos in the cafeteria. She didn't have time to ask all 823 students in her school. Instead, she asked 50 students. She used this information to predict how many tomatoes the cooks should order.

Parker entered a free throw fundraiser at a local high school. He would attempt two hundred free throws and raise money for each free throw made. Before sponsoring him, his aunt asked him how many free throws he would probably make. He decided to shoot 20 free throws and use the information to predict how many free throws he would make in the contest.

Karen and Parker performed experiments. Every experiment can have different **outcomes**. The outcomes of all experiments are the different results you can get.

> **Outcomes**
> Karen's possible outcomes: Yes or No
> Parker's possible outcomes: Make or Miss

The collection of all possible outcomes is called the **sample space**. Use { } to list a sample space.

EXAMPLE 1 Abram had 2 red marbles and 1 white marble in a bag. He chose the white marble. Identify the outcome and show the sample space for this experiment.

SOLUTION

He chose a white marble.   Outcome: white

The possible marbles Abram could have   Sample Space = {red, red, white}
chosen are red, red or white.

Each time an experiment occurs, it is called a **trial**. The more trials you do, the more accurate the estimate will be.

Both Karen and Parker found the **experimental probability** of an event occurring. Experimental probability is the ratio of the number of times an event occurs to the total number of trials.

The notation P( ) is used for probabilities. The probability a person said yes to Karen would be written P(yes). The probability Parker misses the free throw would be written P(miss).

94 Lesson 3.6 ~ Experimental Probability

EXPERIMENTAL PROBABILITY

$$P(\text{event}) = \frac{\text{number of times the outcome occurs}}{\text{total number of trials}}$$

Twenty-four of the 50 students said "yes" to Karen. → $P(\text{yes}) = \frac{24}{50}$ or $\frac{12}{25}$

Parker made 16 free throws out of 20. → $P(\text{make}) = \frac{16}{20}$ or $\frac{4}{5}$

EXAMPLE 2 Nik took a jelly bean out of a bag without looking. He recorded the color. He placed it back in the bag. He did this several times. The results are shown in the table.

Color	Red	Yellow	White	Green
Frequency (number of times chosen)	4	12	8	24

a. Find the total number of trials.
b. Find the experimental probability his next jelly bean will be red.
c. Find the experimental probability his next jelly bean will be green.

SOLUTIONS

a. Find the total of the frequencies. $4 + 12 + 8 + 24 = 48$
   Nik completed 48 trials.

b. $P(\text{red}) = \frac{\text{number of times red is chosen}}{\text{total number of trials}}$ $P(\text{red}) = \frac{4}{48} = \frac{1}{12}$

This is read, "The probability of picking a red is one-twelfth."

c. $P(\text{green}) = \frac{\text{number of times green is chosen}}{\text{total number of trials}}$ $P(\text{green}) = \frac{24}{48} = \frac{1}{2}$

Experimental probabilities are most often written as fractions because you write a fraction to find them. Probabilities can be written as fractions, decimals or percents.

Parker made 16 free throws out of 20. The probability that he will make a free throw shot can be written as a fraction, decimal or percent.

$$16 \text{ out of } 20 = \frac{16}{20} = \frac{4}{5}$$

$$\frac{4}{5} = 0.8$$

$$0.8 = 80\%$$

There is an 80% chance that Parker will make a free throw.

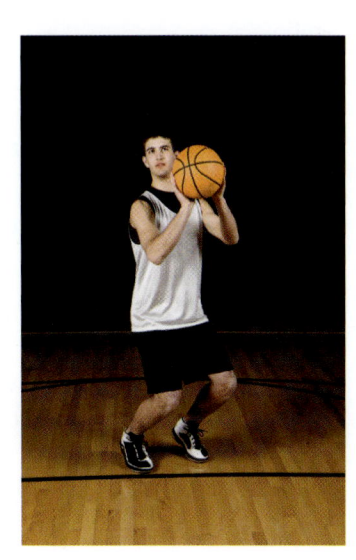

Lesson 3.6 ~ Experimental Probability

EXPLORE!
ROLLING A 3

Step 1: Copy the chart below on your paper. Roll a number cube ten times. Record the number of times each number appears by using a tally mark in the chart.

Number	1	2	3	4	5	6
Frequency (Number of times it occurs)						

    a. How many trials have you done?
    b. Find P(roll a 3).

Step 2: Roll the number cube ten more times. Record the number of times each number appears using the chart from **Step 1**. Continue adding tally marks.
    a. How many total trials have you done?
    b. Find P(roll a 3).

Step 3: Roll the number cube ten more times. Record the number of times each number appears using the chart from **Step 1**. Continue adding tally marks.
    a. How many total trials have you done?
    b. Find P(roll a 3).

Step 4: Roll the number cube ten more times. Record the number of times each number appears using the chart from **Step 1**. Continue adding tally marks.
    a. How many total trials have you done?
    b. Find P(roll a 3).

Step 5: Roll the number cube ten more times. Record the number of times each number appears using the chart from **Step 1**. Continue adding tally marks.
    a. How many total trials have you done?
    b. Find P(roll a 3).

Step 6: Which probability do you think is the most accurate estimate for the probability of rolling a 3 on your next turn? Explain your reasoning.

EXERCISES

Identify each sample space.

1. Joy flips a quarter.

2. Zack rolls a regular six-sided number cube.

3. Peggy picks a marble out of a bag with 2 blue marbles and 1 red marble.

4. Omar picks a card from a deck of cards numbered 1 through 9.

5. Megan has a can of pick-up-sticks. The can contains 1 blue stick, 4 green sticks and 2 red sticks.

6. Bryan can invite one of his three friends, Bob, Tim or Joe to go on a camping trip with him.

7. Terrell put his CD player on random play. It chose the same song 4 times out of the last 10 times it played a song. What is the experimental probability it will play the same song for its next selection?

8. Tom practiced field goals for the football team. He made 14 of his last 20 attempts. What is the experimental probability he will make a field goal on his next attempt?

9. Sharla hit 3 homeruns in 9 at bats during batting practice. Find the experimental probability her next hit will be a homerun.

10. Lynette went surfing at Torrey Pines Beach. She caught 4 of the 20 possible waves to catch. What is the experimental probability she will catch the next possible wave?

11. Penny asked people if they would vote for her for student council. The results were: yes, no, yes, yes, yes, yes, yes, yes, no, yes. Penny said that means the experimental probability the next person will say no is $\frac{1}{5}$. Is she correct? Explain your reasoning.

12. Use any coin for this experiment.
   a. Flip the coin 10 times. Record the number of heads and the number of tails. Find the experimental probability of the coin landing heads.

Heads	Tails

   b. Flip the coin 10 more times. Add the number of heads and the number of tails to the original chart. How many trials have you done? Find the experimental probability of the coin landing heads.
   c. Repeat this process one more time. Find the experimental probability of the coin landing heads.
   d. After 30 coin flips, how many times would you expect to get heads? Is this what you found during your experiment? Why or why not?

13. Emmit recorded the number of people who bought each type of vehicle on his used car lot in one day.

Vehicle	Truck	Van	SUV	Car
Frequency	2	4	3	6

   a. Find the experimental probability the next customer who buys a vehicle will buy each type of car.

Vehicle	Truck	Van	SUV	Car
Experimental Probability				

   b. Based on these probabilities, which vehicle is most likely to be bought next?

14. Tamara used the spinner to the right. She recorded the number of times she landed on each color.

Color	Red	Yellow	Blue
Frequency	5	4	11

   a. Find the experimental probability of spinning yellow on the next turn. Write the probability as a fraction, decimal and percent.
   b. Find the experimental probability of spinning blue on the next turn. Write the probability as a fraction, decimal and percent.

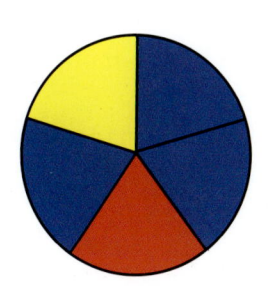

Lesson 3.6 ~ Experimental Probability

15. Madison randomly chose a card from a deck of playing cards. She put the card back and chose again. Each time she recorded the suit of the card: diamonds, hearts, spades or clubs. Her results are shown below.

Suit	♦ Diamonds	♥ Hearts	♠ Spades	♣ Clubs
Frequency	8	15	7	10

  a. Find the total number of trials.
  b. Find the experimental probability of picking a Diamond on her next turn. Write the probability as a fraction, decimal and percent.
  c. Find the experimental probability of picking a Heart on her next turn. Write the probability as a fraction, decimal and percent.
  d. Find the experimental probability of picking a Spade on her next turn. Write the probability as a fraction, decimal and percent.
  e. Find the experimental probability of picking a Club on her next turn. Write the probability as a fraction, decimal and percent.
  f. Based on her experiment, which suit is most likely to appear on her next turn?

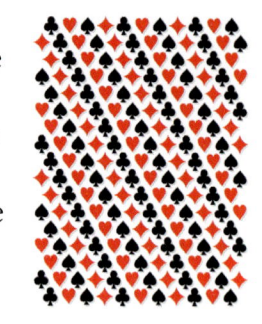

REVIEW

Order the probabilities from least to greatest.

16. 45%, $\frac{2}{5}$, 0.48

17. 0.2, $\frac{2}{11}$, 23%

18. 50%, 0.05, $\frac{5}{20}$

19. Shane bought a printer in California for $300. The city charged an additional 8% sales tax. How much did Shane pay for the printer, including sales tax? Show all work necessary to justify your answer.

20. Hiro bought lunch at the local deli. His sandwich and drink cost $12.00. He left a 15% tip for the server. How much did Hiro pay for lunch, including the tip? Use mathematics to justify your answer.

TIC-TAC-TOE ~ RAINFALL

Look in an almanac to find the average number of days it rains each month in Seattle, Washington.

Step 1: Record the average number of days it rains. Find the probability it will rain on a given day each month.

Step 2: Determine whether it is *impossible, unlikely, equally likely, likely* or *certain* to rain on a given day each month.

Step 3: Which month has the highest chance of rain on any given day?

Look in an almanac to find the average number of days it rains in New York, Miami and your city. Repeat **Steps 1-3** for each place. (If your city has already been selected, choose another city in the United States.)

Tic-Tac-Toe ~ Eye Color

Survey 25 students to ask the color of their eyes. Copy the chart below and record the colors as brown, blue, hazel or green.

	Brown	Blue	Hazel	Green
Frequency (number of students)				

Step 1: Find the experimental probability for each eye color. Write the probability as a fraction, decimal and percent.

Step 2: Which color is most common among the students you asked?

Step 3: Do you think you would have similar results if you asked 50 students instead of 25? Ask 25 more students. Add their data to the chart above.

Step 4: Find the new experimental probability for each eye color after asking 50 students. Write the probability as a fraction, decimal and percent.

Step 5: Did any of the probabilities change? Explain why or why not using complete sentences.

Step 6: You have to predict the actual percent of students with each eye color at your school. Would you rather use the probabilities after asking 25 students or 50 students? Explain your reasoning.

Tic-Tac-Toe ~ Jelly Beans

Cedar has a bag of jelly beans. Inside the bag are 10 red, 12 white, 8 yellow and 5 green jelly beans.

Cedar grabs a jelly bean and puts it back after each turn.

1. Find P(red)   **2.** Find P(white)   **3.** Find P(yellow)   **4.** Find P(green)

Now he grabs a jelly bean and eats it before choosing the next jelly bean.

5. Find P(first jelly bean is red).
6. Given that Cedar grabbed a red jelly bean and ate it on the first pick, find the probability that the second jelly bean will be red. Explain how you know your answer is correct.
7. Given that the first two jelly beans Cedar grabbed and ate were red, find P(white) on his third grab.
8. Cedar has eaten two red and one white jelly bean. Find P(yellow) on his next grab.
9. Explain the difference between finding probabilities if everything is replaced after a turn and finding probabilities if the item remains out after every turn. Use complete sentences.

Lesson 3.6 ~ Experimental Probability

THEORETICAL PROBABILITY

LESSON 3.7

Find and interpret the theoretical probability of an event.

EXPLORE! SUM OF TWO NUMBER CUBES

Step 1: Copy and complete the chart below. It shows the possible outcomes of one number cube across the top, and a second down the left column. The corresponding sums are shown in the table. Fill in the missing sums.

+	1	2	3	4	5	6
1	2	3	4	5	6	7
2	3	4	5	6	7	8
3	4	5	6	7	8	9
4	5	6	7	8	9	10
5	6	7	8	9	10	11
6	7	8	9	10	11	12

a. How many sums are shown in the chart?

b. Find and record the frequency of each sum (the number of times each sum appears).

Sum	2	3	4	5	6	7	8	9	10	11	12
Frequency	1	2	3	4	5	6	5	4	3	2	1

Step 2: Copy and complete the table below to find the probability of rolling each sum on this chart.

Sum	2	3	4	5	6	7	8	9	10	11	12
Probability											

For example, to find the probability of rolling a 7, find $\frac{\text{number of ways to get a sum of 7}}{\text{number of outcomes possible}} = \frac{6}{36} = \frac{1}{6}$.

Step 3: Which sum is most likely to occur?

Step 4: Roll two number cubes 36 times. Copy the table and record the sums with tally marks.

Sum	2	3	4	5	6	7	8	9	10	11	12
Frequency											

Step 5: Find the experimental probability of rolling a sum of 7 on your next turn. Does your experimental probability match the probability in **Step 2**? Explain your reasoning.

You found probabilities by doing experiments in the last lesson. Sometimes the outcomes did not have the same probability as you would have expected. When flipping a coin, you would expect to get as many heads as tails. This is called theoretical probability. **Theoretical probability** is the expected ratio of favorable outcomes (wanted outcomes) to the number of possible outcomes.

> **THEORETICAL PROBABILITY**
>
> $$P(\text{event}) = \frac{\text{number of favorable outcomes}}{\text{number of possible outcomes}}$$

EXAMPLE 1

a. Find P(tails) when flipping a coin.
b. Find P(3 or 5) when rolling one number cube.
c. Find P(not 6) when rolling one number cube.

SOLUTIONS

a. $P(\text{tails}) = \dfrac{\text{number of favorable outcomes (tails)}}{\text{number of possible outcomes (heads or tails)}} = \dfrac{1}{2}$

b. $P(3 \text{ or } 5) = \dfrac{\text{number of favorable outcomes (3, 5)}}{\text{number of possible outcomes (1, 2, 3, 4, 5, 6)}} = \dfrac{2}{6} = \dfrac{1}{3}$

c. $P(\text{not } 6) = \dfrac{\text{number of favorable outcomes (1, 2, 3, 4, 5)}}{\text{number of possible outcomes (1, 2, 3, 4, 5, 6)}} = \dfrac{5}{6}$

You will always get a head or tail when you flip a coin, so P(heads or tails) = 1. Also, when rolling a number cube, P(1, 2, 3, 4, 5 or 6) = 1 since those are the only outcomes possible.

Another way to find P(not 6) when rolling a number cube is determining P(6) and subtracting the answer from 1.

$P(\text{not } 6) = 1 - P(6) = 1 - \dfrac{1}{6} = \dfrac{5}{6}$ This is the same as the answer in **Example 1c**.

P(not 6) and P(6) are called **complements** because together they contain all possible outcomes with no overlap. The sum of complements is always 1.

EXAMPLE 2

Each letter from the word MATHEMATICS is written on a separate card. A card is chosen at random. Find the probability of each event.

a. P(C)   b. P(M)   c. P(not M)   d. P(vowel)

SOLUTIONS

a. $P(C) = \dfrac{1 \text{ C}}{11 \text{ letters possible}} = \dfrac{1}{11}$

b. $P(M) = \dfrac{2 \text{ Ms}}{11 \text{ letters possible}} = \dfrac{2}{11}$

c. $P(\text{not } M) = 1 - P(M) = 1 - \dfrac{2}{11} = \dfrac{9}{11}$

d. $P(\text{vowel}) = \dfrac{A, E, A, I}{11 \text{ letters possible}} = \dfrac{4}{11}$

Lesson 3.7 ~ Theoretical Probability

EXERCISES

Find each probability for one roll of a number cube. Write each answer as a fraction in simplest form.

1. P(2) **2.** P(3, 4 or 5) **3.** P(even number)

4. P(0) **5.** P(not 2) **6.** P(less than 7)

Each letter from the word PROBABILITY is written on a separate card. A card is chosen at random. Find the probability of each event. Write each answer as a fraction in simplest form.

7. P(P) **8.** P(not P) **9.** P(vowel)

10. P(consonant) **11.** P(B or I) **12.** P(Q)

You play a game with 20 cards numbered 1 through 20. The cards are shuffled and one is picked at random from the complete deck. Find the probability of each event as a fraction in simplest form.

13. P(4) **14.** P(30) **15.** P(odd number)

Use the spinner to find each probability. Write each probability as a fraction, decimal and percent.

16. P(blue)

17. P(even number)

18. P(red)

19. P(not green)

20. P(9)

21. Write two different probabilities from the spinner that have the same value. What is the value?

22. Jermaine created his own card game. The probability of getting a blue card was $\frac{1}{3}$. Find P(not blue).

23. Bena has seven cats. Three of the cats are black. One of the cats was chosen at random. What is the probability that it is not black?

24. The probability of landing on the purple region of a spinner is $\frac{4}{11}$. Find P(not purple).

25. Todd had a bag of marbles containing 5 blue, 3 green and 2 yellow marbles. He reached into the bag and pulled out a yellow marble. He did not return the yellow marble to the bag.
   a. Find the probability the second marble drawn will also be yellow.
   b. Find the probability the second marble drawn will be green.

26. A game required each player to roll two number cubes and find their sum. Complete the chart below to find all possible sums. The top row and left column are the possible numbers rolled on each number cube. The inside of the chart shows the corresponding sum. Find:

    a. P(4)
    b. P(not 4)
    c. P(2 or 12)
    d. P(even)
    e. P(0)
    f. P(less than 13)
    g. Carrie rolls the number cubes 15 times trying to get a sum of 7 to win the game. She never gets a sum of 7. Do you think the number cubes are fair? Explain your reasoning.

Number Cube Sums

+	1	2	3	4	5	6
1	2	3	4	5	6	7
2	3	4	5	6	7	8
3						
4						
5						
6						

27. Copy the spinner on your paper. Color it so each probability is true when a person spins it.

    a. P(green) = $\frac{1}{2}$

    b. P(blue) = $\frac{1}{6}$

    c. P(red) = $\frac{1}{12}$

    d. P(yellow) = $\frac{1}{4}$

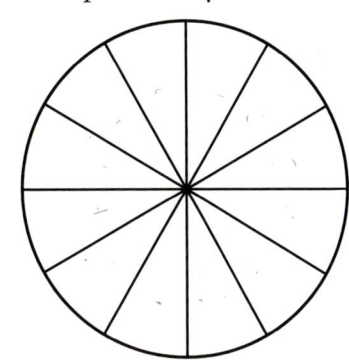

REVIEW

28. Addison rolled a number cube 20 times. It landed on 3 four times.
    a. Find the experimental probability she will roll a 3 on her next roll.
    b. Find the theoretical probability she will roll a 3 on her next roll.

29. Sai had a bag of marbles: 5 blue, 8 red, 2 yellow and 5 green. He chose one marble from the bag.
    a. Find the theoretical probability Sai will choose a blue marble.
    b. After choosing and returning marbles 10 times, Sai chose a blue marble 6 times. Find the experimental probability Sai will choose a blue marble on his next pick.

30. There is a 30% chance of rain. Which phrase best describes the chance of rain?
    impossible, unlikely, equally likely, likely, certain

Solve each percent problem.

31. 40% of 800 is _____

32. 225% of 100 is _____

33. 15% of 30 is _____

Tic-Tac-Toe ~ Number Cube Differences

Step 1: Copy and complete the chart below which shows one number cube across the top, the second down the left column and the corresponding differences in the middle. Fill in the missing differences. Always subtract the smaller number from the larger number.

-	1	2	3	4	5	6
1	0	1	2	3	4	5
2	1	0	1	2	3	4
3						
4						
5						
6						

a. How many differences are shown in the chart?

b. Find the frequency of each difference (the number of times each difference appears).

Difference	0	1	2	3	4	5
Frequency						

Step 2: Copy and complete the chart below. Find the theoretical probability of rolling each difference on this chart.

Example: Find the theoretical probability of rolling a difference of 1.

$$\frac{\text{number of ways to get a difference of 1}}{\text{total number of outcomes possible}} = \frac{10}{36} = \frac{5}{18}$$

Difference	0	1	2	3	4	5
Probability as a fraction		$\frac{5}{18}$				
Probability as a decimal						
Probability as a percent						

Step 3: Which difference is most likely to occur?

Step 4: Roll two number cubes 36 times. Copy the table and record the differences with tally marks.

Difference	0	1	2	3	4	5
Frequency						
Experimental probability						

Step 5: a. Find the experimental probability of rolling each difference.

b. How does your experimental probability compare to the theoretical probability in **Step 2**? Explain why there may be similarities or differences.

GEOMETRIC PROBABILITY

LESSON 3.8

Find the geometric probability of an event.

EXPLORE!   WHAT ARE MY CHANCES OF WINNING?

Step 1: Draw a 6 inch by 6 inch square on a sheet of paper. Draw a diagonal line and shade one side as shown. This is the game board.

Step 2: What percent of the square is shaded? Write your answer as a fraction, decimal and a percent.

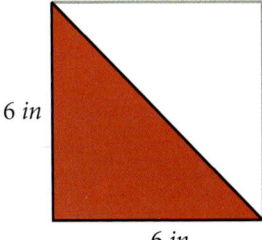

Step 3: Hold a bean above the center of the square game board. Drop the bean so it falls on the game board. If it does not fall on the game board, drop it again until it does. Only count trials where the bean lands on the square. Record whether or not it fell in the shaded area of the square. When the bean lands in the shaded area record a "win". If it falls in the unshaded area record a "loss". Drop the bean on the board at least 10 times.

Win (Shaded Area)	Loss

Step 4: Find the experimental probability that a bean will land in the shaded area.

$$P(\text{lands in shaded area}) = \frac{\text{number of beans in the shaded area}}{\text{number of beans dropped}}$$

Step 5: Find the theoretical probability a bean will land in the shaded area by answering these questions.

$$P(\text{lands in the shaded area}) = \frac{\text{number of favorable outcomes}}{\text{number of possible outcomes}} = \frac{\text{area of the shaded region}}{\text{area of the entire square}}$$

    a. Find the area of the shaded region.
    b. Find the area of the entire square.
    c. Find the theoretical probability the bean lands in the shaded area. Write your answer as a fraction, decimal and percent.

Step 6: How do your answers in **Step 2** and **Step 5c** compare? Explain your reasoning.

Step 7: Create a game board that has a $\frac{1}{4}$ chance of winning. Explain how you know you have a 25% chance of winning.

Lesson 3.8 ~ Geometric Probability   105

Finding the theoretical probability of winning the game in the **Explore!** is an example of geometric probability. **Geometric probability** is probability based on length and area. This means it is a ratio comparing two lengths or a ratio comparing two areas.

EXAMPLE 1

William just finished painting 40 feet of fence along the side of his house. He still has 50 feet left to paint. A bird landed on the fence. What is the probability the bird landed on the painted part of the fence?

SOLUTION

Find the ratio of the painted length of fence to the total length of fence.

$$P(\text{bird lands on painted part}) = \frac{\text{length of painted part of fence}}{\text{total length of fence}}$$

The painted length is 40 feet.
The unpainted length is 50 feet.

| 40 ft | 50 ft |

Find the entire length of the fence.   40 + 50 = 90 feet

Find the probability. $P(\text{bird lands on painted part}) = \frac{40 \text{ feet}}{90 \text{ feet}} = \frac{4}{9}$

The probability the bird landed on the painted part of the fence is $\frac{4}{9}$.

EXAMPLE 2

What is the probability that a dart randomly hitting the board below will land in the red triangle?

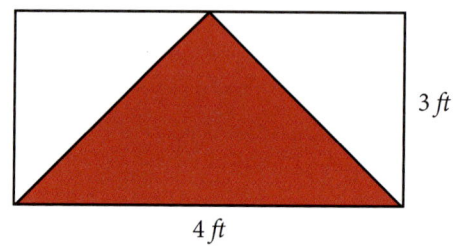

3 ft
4 ft

SOLUTION

$$P(\text{dart lands in the red triangle}) = \frac{\text{area of red triangle}}{\text{area of dart board (rectangle)}}$$

The base of the red triangle is 4 and its height is 3 (same as the rectangle).

Find the area of the red triangle. $\frac{1}{2} \times 4 \times 3 = 6 \text{ ft}^2$

Find the area of the rectangle. $4 \times 3 = 12 \text{ ft}^2$

Find the probability. $P(\text{dart lands in the red triangle})$
$= \frac{6 \text{ ft}^2}{12 \text{ ft}^2} = \frac{1}{2}$

The probability a dart will land in the red triangle is $\frac{1}{2}$.

Lesson 3.8 ~ Geometric Probability

EXERCISES

Find the perimeter and area of each shape below. Show all work necessary to justify your answer.

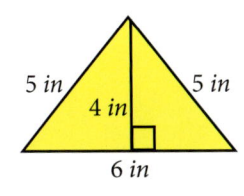

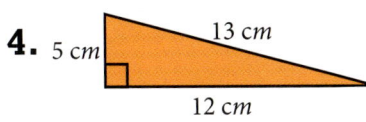

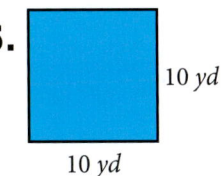

1. Rectangle: 4 in by 7 in

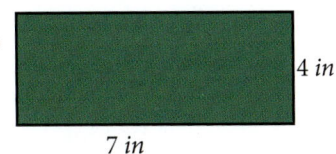

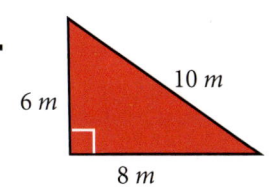

2. Triangle: sides 5 in, 5 in; base 6 in; height 4 in

3. Right triangle: legs 6 m and 8 m; hypotenuse 10 m

4. Right triangle: legs 5 cm and 12 cm; hypotenuse 13 cm

5. Square: 10 yd by 10 yd

6. Rectangle: 5 ft by 20 ft

7. Kevin painted 32 feet of railing. He still has 64 feet of railing left to paint. Kelsey didn't realize some of the railing had been painted and leaned against it. Find the probability that she leaned against the freshly painted part of the railing.

8. Evan's father set up a surround-sound system. The kit came with a piece of red and a piece of blue wire. One of the wires had been accidently clipped and could not be used. Find the probability of the clip occurring on the red wire. Complete each part below.

   Red wire: ―――――――――――――
   Blue wire: ――――――――――――――――――

   a. Measure the red wire in centimeters. Record the measurement.
   b. Measure the blue wire in centimeters. Record the measurement.
   c. Find P(clip on red wire) = $\dfrac{\text{length of red wire}}{\text{total length of wires}}$

Use each figure below to find the geometric probability that a dart randomly landing on each board will land in the shaded part. Write each answer as a simplified fraction.

9. Rectangle divided into four parts: top row 2 in tall (shaded left 4 in, unshaded right 4 in), bottom row 2 in tall (unshaded left 4 in, shaded right 4 in).

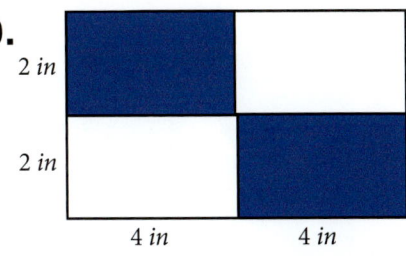

10. Triangle with height segments 3 m, and 3 m; base 1 m + 2 m + 1 m, with a shaded square of side 2 m inside.

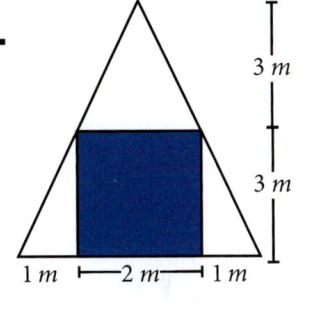

11. Right triangle with vertical leg 1 m + 4 m and horizontal leg 4 m + 16 m; shaded square of side 4 m.

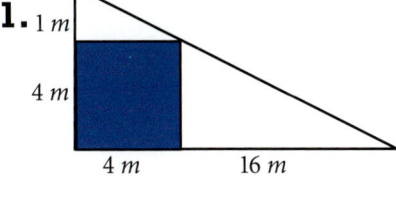

12. Find the probability that the dart will not land in the shaded part of each dart board in **Exercises 9-11**.

Lesson 3.8 ~ Geometric Probability   107

Use each figure below to find the geometric probability that a dart landing on each board will land in the shaded part. Write each answer as a simplified fraction.

13.

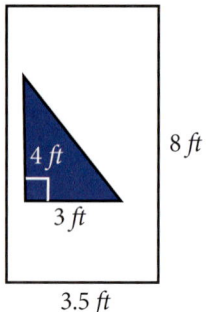

14.

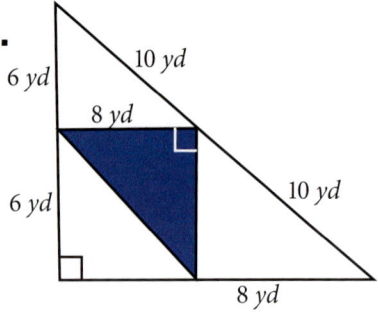

15.

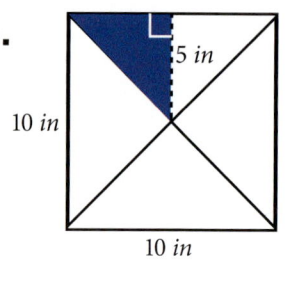

16. Find the probability that the dart will not land in the shaded part of each dart board in **Exercises 13-15**.

REVIEW

17. The Armstrong family travels 140 miles in 4 hours. At this rate, how far will they travel in 7 hours?

18. The basketball team held a fundraiser. Pasha sold 20 raffle tickets for a total of $30. Nancy sold 50 raffle tickets. How much money did Nancy raise? Show all work necessary to justify your answer.

19. Jeannie bought 6 pounds of pears for $6.42. Judy bought 4 pounds of pears. How much did Judy spend on pears?

Solve each percent problem.

20. 25% of 40 is _____

21. 15% of 24 is _____

22. _____ is 50% of 34

23. _____ is 12% of 50

24. 62% of 100 is _____

25. _____ is 90% of 10

Tic-Tac-Toe ~ Carnival Game

To win a carnival game you have to throw a dart that lands in the yellow part of the star. Your dart hits the board. What is the probability it will hit the yellow part of the star?

The base of each triangle is 12.5% of the length of the side of the dartboard. The height of each triangle is 25% of the length of the side of the dartboard.

108   Lesson 3.8 ~ Geometric Probability

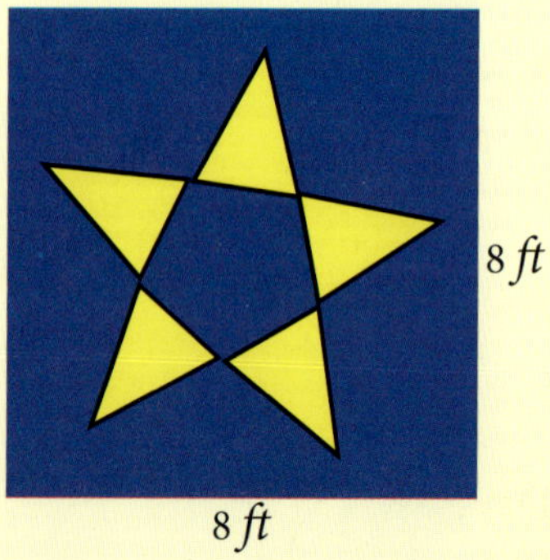

REVIEW

BLOCK 3

Vocabulary

complements
discount
experimental probability
geometric probability

outcomes
percent
probability

sales tax
sample space
theoretical probability
trial

- Write percents as fractions and decimals.
- Write fractions and decimals as percents.
- Find the percent of a number.
- Solve problems involving discounts, tips and tax.
- Recognize probabilities that are unlikely and likely.
- Express probabilities as fractions, decimals and percents.
- Find and interpret the experimental probability of an event.
- Find and interpret the theoretical probability of an event.
- Find the geometric probability of an event.

Lesson 3.1 ~ Introducing Percents

For each shaded grid, write the ratio of the shaded squares to 100 as a fraction. Write the percent of squares shaded as a number with the % sign.

1.

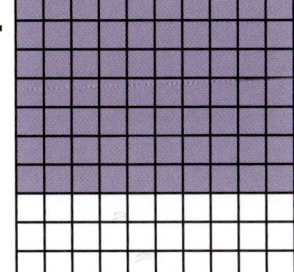

2.

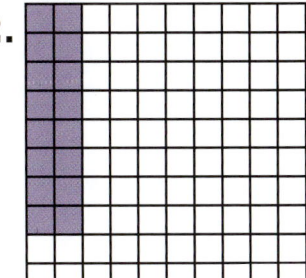

3.

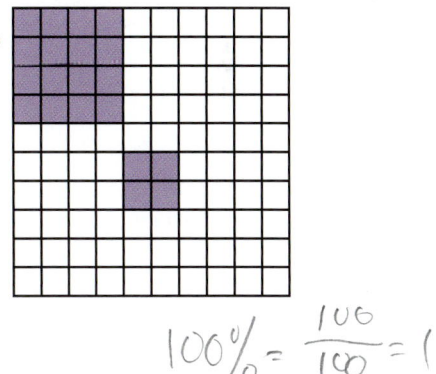

Write each percent as a fraction in simplest form.

4. 10%

5. 25%

6. 400%

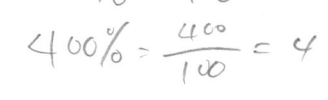

Write each percent as a decimal.

7. 35%

8. 70%

9. 1%

10. Marcia found her favorite jeans on sale for 20% off the original price. What percent of the price is she going to pay for the jeans?

11. David and Leon went to the store and found a sale on video games. David said the games were 80% of their original price. Leon said the games were 20% off their original price. Explain how both David and Leon can be correct.

Block 3 ~ Review   109

Lesson 3.2 ~ Percents, Decimals and Fractions

12. Write each value from the list below that is equivalent to three-fifths.

35% 0.6 $\frac{3}{5}$ 30% 3.5 $\frac{6}{100}$ 60%

Write the shaded part of each shape as a fraction in simplest form, as a decimal and as a percent.

13.

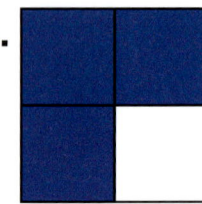

14.

15.

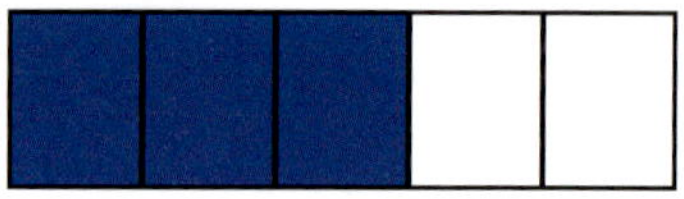

Write each decimal as a percent.

16. 0.05

17. 0.8

18. 1.5

Write each fraction as a percent.

19. $\frac{1}{4}$

20. $\frac{9}{10}$

21. $\frac{2}{3}$

Order the numbers from least to greatest.

22. 45%, 0.35, $\frac{2}{5}$

23. $\frac{1}{3}$, 30%, 0.34

24. 0.2, 2%, $\frac{2}{3}$

25. Kahlia said the decimal 1.5 was equivalent to 15%. Is she correct? Explain your reasoning.

Lesson 3.3 ~ Percent of a Number

26. Lyle has 40% of his family's 300 baseball cards. Justin has 60% of his family's 200 baseball cards. Who has more baseball cards? Use mathematics to justify your answer.

Solve each percent problem.

27. 25% of 80 is _____

28. _____ is 10% of 72

29. 50% of 200 is _____

30. 15% of 30 is _____

31. 80% of 150 is _____

32. _____ is $33\frac{1}{3}$% of 90

33. Find 45% of $1.00.

34. Find 24% of $100.

35. Find 75% of 1 meter.

36. Draw 25% of the line.

37. Draw 80% of the line.

Lesson 3.4 ~ Percent Applications

38. Trent bought a new flash drive. The flash drive was originally $25.00, but was on sale for 20% off the original price. How much was the discount on the flash drive?

39. Tex went to the store to buy a new laptop. He found a laptop he liked that was discounted 10%. The original price was $750.00. How much was the laptop after the discount? Show all work necessary to justify your answer.

40. George and Will went to a restaurant for lunch. The bill was $32.00. They left a 15% tip. Find the amount they left for the tip.

41. Alden and Jenny went out for dinner. The bill for the meal was $48.00. Jenny wanted to leave a 15% tip. Find the total cost of the meal, including the tip. Show all work necessary to justify your answer.

42. Shayna bought a new digital camera while on vacation in Florida. She paid an additional 6% sales tax on the $275 camera. How much sales tax did Shayna pay on the camera?

43. Hector bought a new suit in California. It had a price tag of $350.00. He paid an additional 8% in sales tax. How much did Hector pay for the suit, including the sales tax? Show all work necessary to justify your answer.

Lesson 3.5 ~ Introduction to Probability

Determine whether each event is *impossible, unlikely, equally likely, likely* or *certain*.

44. You roll a 2 on a regular number cube.

45. You flip a coin and get heads.

46. You roll a number less than 7 on a regular number cube.

47. The teacher assigns the date December 32nd to turn in your homework.

48. You pick a green marble from a bag where 80% of the marbles in the bag are green.

49. Bill had a bag of jelly beans with several colors. The probability that the next jelly bean he picks will be pink is 75%.
    a. Write this probability as a fraction and as a decimal.
    b. Describe what the contents of the bag of jelly beans may look like based on this probability.

50. The probability that Gil will be asked to shoot the last shot in the basketball game is 0.2.
    a. Write this probability as a fraction and as a percent.
    b. Do you think Gil will be asked to shoot the last shot? Explain your reasoning.

51. Fred polled 40 people and asked them which sport they preferred to watch on TV. Fifty-five percent chose football, 0.15 chose baseball and $\frac{3}{10}$ chose basketball.
  a. Which sport did most people prefer to watch? Explain your reasoning.
  b. Find the number of people out of 40 who preferred each sport.

Lesson 3.6 ~ Experimental Probability

Find the sample space for each experiment in Exercises 52-54.

52. Marlon flips a quarter.

53. Patty rolls a regular number cube.

54. John chooses a marble out of a bag containing 2 red marbles, 3 blue marbles and 1 white marble.

55. Ted practiced kicking goals from the 10-yard line before his soccer game. Seven of the 20 kicks were on goal. What is the experimental probability his shot will be on goal if he shoots from the 10-yard line in the game?

56. Daryl threw 3 darts at a dart board. Two of the 3 darts hit the bullseye. Find the experimental probability his next dart will hit the bullseye.

57. Tito asked 25 students which type of video game they preferred to play. The results are below.

Type of Video Game	Adventure	Action	Racing
Frequency (number of students)	10	6	9

  a. Find the experimental probability the next person he asks will choose adventure games.
  b. Find the experimental probability the next person he asks will choose action games.
  c. Find the experimental probability the next person he asks will choose racing games.
  d. Find the experimental probability the next person he asks will not choose racing games.

58. After rolling a number cube 24 times, Lance found the experimental probability he would roll a 6 on his next turn was $\frac{1}{3}$. How many times had he rolled a 6 on his previous turns? Use mathematics to justify your answer.

59. Lucy says the probability she will roll a "2" on a number cube is $\frac{1}{2}$ because there is one "2" on the number cube. Is she correct? Explain your reasoning.

Lesson 3.7 ~ Theoretical Probability

Find each probability for one roll of a number cube.

60. P(8)   **61.** P(3 or 5)   **62.** P(odd number)   **63.** P(6)

Each letter from the word FOOTBALL is written on a separate card. A card is chosen at random. Find the probability of each event. Write each answer as a fraction in simplest form.

64. P(B)

65. P(L)

66. P(vowel)

67. P(consonant)

68. Two out of five candies in a bag are yellow. If a candy is chosen at random, find P(not yellow).

69. The probability that someone's birthday is January 1st is $\frac{1}{365}$. Find P(not born on January 1st). Explain your reasoning.

70. Bethany's bag of marbles holds 2 blue, 5 green and 3 yellow marbles. She takes one marble out of the bag.
   a. Find the theoretical probability that the marble will be green.
   b. Find the theoretical probability that the marble will be blue.
   c. Find the theoretical probability that the marble will not be blue.

Lesson 3.8 ~ Geometric Probability

Find the perimeter and area of each shape below. Show all work necessary to justify your answer.

71. SQUARE 20 in

72. Triangle: 29 in, 29 in, 21 in (height), 40 in (base)

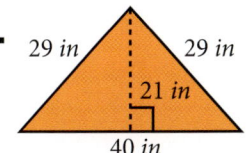

73. Right triangle: 9 cm, 15 cm, 12 cm

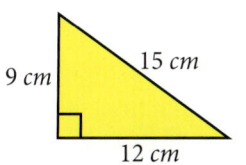

74. Mary put in a new sprinkler system. She used 24 feet of gray pipe and 12 feet of white pipe along the front yard. The next morning she noticed there was a leak somewhere in the front yard pipes. Find the probability that the leak is in the gray pipe. Show all work necessary to justify your answer.

Find the geometric probability that a dart landing on each board will land on the shaded part. Write each answer as a simplified fraction.

75. Square divided into 4 parts, 6 in × 6 in each, top-left shaded.

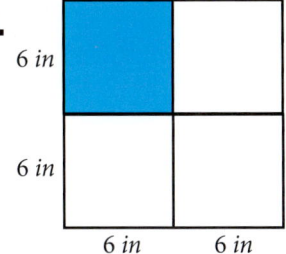

76. Triangle with 4 cm, 2 cm, 4 cm, 4 cm, 4 cm markings; top portion shaded.

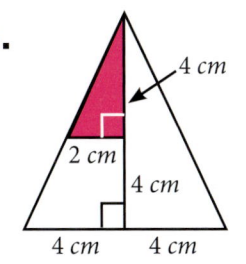

77. Triangle with 15 m, 10 m, 10 m, 15 m markings; right portion shaded green.

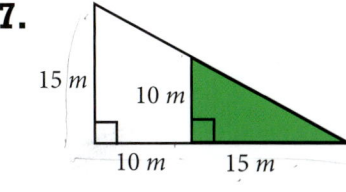

78. Find the probability that the dart will <u>not</u> land on the shaded part of each dart board in **Exercises 75–77**.

CAREER FOCUS

Tom
Commercial Truck Driver

I am a commercial truck driver. I am responsible for delivering products safely to my customers. I must make sure products are delivered on time. Commercial truck drivers have many different types of routes. I drive the same route each day. Some drivers have routes that are different each day. Other drivers may be away from home for many days at a time.

I use math in my job to calculate the mileage I drive each day and for each trip. I maintain a journal of all my hours for the day including stops, breaks and time off. I also have to calculate weights to make sure my truck is safe and that I am following all the rules of the road.

Commercial truck drivers must have a special driver's license. They usually get this after attending a trade school. The training takes 2 to 3 months. There are certain tests to pass at the end of the training. Another option to get a license is being hired by a trucking company which provides the training and the trucks for drivers to pass the tests.

An average starting salary for a truck driver is $25,000 – $35,000 per year. The range of salaries varies depending on the company you work for and the type of driving you do. Some commercial truck drivers make up to $70,000 per year.

I like the independence of my profession. I get to work outdoors. I also get to see beautiful country. Commercial truck driving offers good pay, benefits and security. It is a job that I enjoy.

CORE FOCUS ON RATIOS, RATES & STATISTICS
BLOCK 4 ~ STATISTICS

LESSON 4.1	INTRODUCTION TO STATISTICS	117
	EXPLORE! A QUESTION OF STATISTICS	
LESSON 4.2	MEASURES OF CENTER	122
	EXPLORE! COUNTING PETS	
LESSON 4.3	DOT PLOTS	128
	EXPLORE! HOW TALL?	
LESSON 4.4	HISTOGRAMS	134
LESSON 4.5	BOX-AND-WHISKER PLOTS	141
LESSON 4.6	ANALYZING STATISTICS	148
	EXPLORE! WHAT'S THE "MEAN"ING?	
LESSON 4.7	MEAN ABSOLUTE DEVIATION	154
	EXPLORE! MERCURY'S RISING	
REVIEW	BLOCK 4 ~ STATISTICS	160

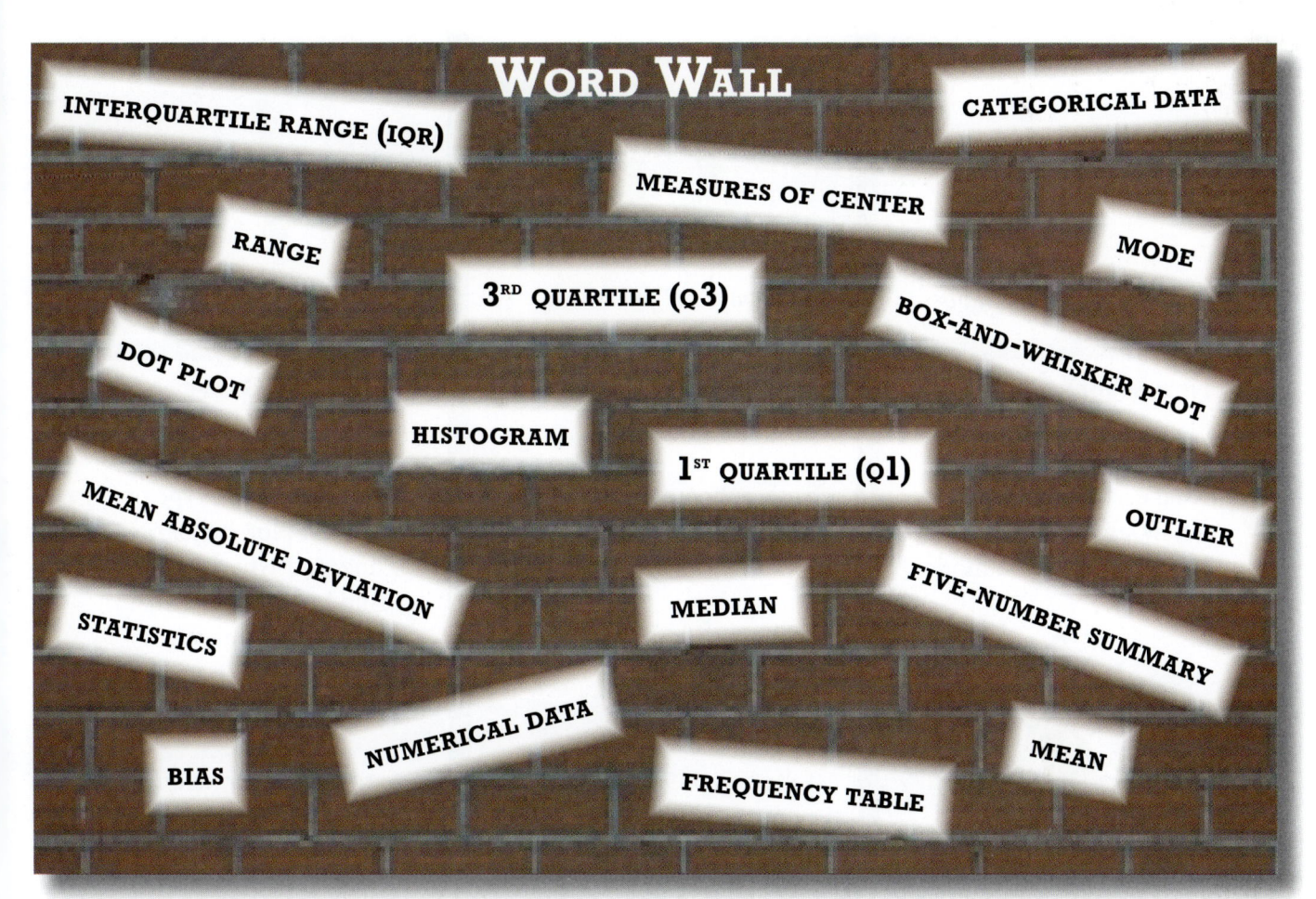

WORD WALL

- INTERQUARTILE RANGE (IQR)
- CATEGORICAL DATA
- MEASURES OF CENTER
- RANGE
- MODE
- 3RD QUARTILE (Q3)
- BOX-AND-WHISKER PLOT
- DOT PLOT
- HISTOGRAM
- 1ST QUARTILE (Q1)
- OUTLIER
- MEAN ABSOLUTE DEVIATION
- MEDIAN
- FIVE-NUMBER SUMMARY
- STATISTICS
- NUMERICAL DATA
- MEAN
- BIAS
- FREQUENCY TABLE

BLOCK 4 ~ STATISTICS
TIC-TAC-TOE

MAKE IT COMPLETE!	**DOUBLE THE WHISKERS**	**EX"SKEW"SE MY LEANING**
Make complete data sets using information and clues that are given. *See page 133 for details.*	Learn to create and interpret parallel box-and-whisker plots. *See page 165 for details.*	Learn about and make a poster describing "skew" in a variety of graphs and situations. *See page 140 for details.*
HOW FAR IS TOO FAR?	**PREDICTION TIME!**	**SHIFTING DATA**
Use the "IQR Method" to determine outliers for data sets. *See page 153 for details.*	Use your knowledge of data graphs and rates to make predictions. *See page 147 for details.*	Learn how changing values in a data set affects the mean and median. *See page 127 for details.*
NEW STUDENT BROCHURE	**THE BIG AND SMALL OF IT**	**WHICH GRAPH SHOULD I USE?**
Make a brochure to help a new student learn about mean, median and mode. *See page 133 for details.*	Explore how outliers affect large and small data sets. *See page 159 for details.*	Determine which type of graph (dot plot, histogram or box-and-whisker plot) would be best to use in a specific situation. *See page 158 for details.*

INTRODUCTION TO STATISTICS

LESSON 4.1

Identify and write statistical questions and the type of data that will result.

On average, how many people visit Disneyland each month? How many people are in a typical class at your school? What is the favorite ice cream flavor of students in your class? These are all examples of statistical questions.

Statistics is the process of collecting, displaying and analyzing a set of data. Consumers may use the data to decide which products to buy. Businesses use data to make important financial decisions. A good statistical question is one in which the data will vary.

A Statistical Question	Not a Statistical Question
How old are the students at your school?	How old is the oldest student at your school?
What grades did students earn in the class?	What is the highest grade a student can earn in this class?
How many people went to each movie at the theater?	What movies are showing at this theater?

Data can either be categorical or numerical. **Categorical data** is data collected in the form of words. For example, if you collected data on your friends' favorite food, they would respond with specific words (pizza, tacos, hamburgers, etc.). **Numerical data** is data that is collected in the form of numbers. For example, if you asked your friends how many CDs they owned, they would respond with a number (5, 32, etc.).

EXPLORE!   A QUESTION OF STATISTICS

Step 1: For each question below, determine if it is a statistical question or not. Make a list of each question that is a statistical question. There should be 6 in your list.

- C   A. What type of pets do students have?
- C   B. What is the fastest land mammal?
- C   C. What is each student's favorite TV show?
- N   D. How much does an airplane ticket cost?
- N   E. How many books are checked out daily from the library?
- N   F. How much money do doctors make?
- N   G. How many teams are in the NFL?
- C   H. Which movie made the most money in 2012?
- C   I. How did students get to school today?
- N   J. How many books were checked out yesterday from the library?

Step 2: For each question in your list from **Step 1**, decide whether the question would give categorical data or numerical data. Write "Categorical" or "Numerical" next to each question.

Step 3: Create two more statistical questions. Identify whether each question would give categorical or numerical data.

Lesson 4.1 ~ Introduction to Statistics   117

EXAMPLE 1

Mrs. Crenshaw asks her 6th grade students the following questions:
    How old are you in years?   *How old are you in months?*
Which question will give data that is more varied? Explain your reasoning.

SOLUTION

The second question will give more varied answers. Since Mrs. Crenshaw's students may all be close to the same age in years, there may be little difference in students' answers (11- or 12-years old). Because there are 12 months in a year, asking for students' ages in months could give far more varied answers. For example, 11 years and 3 months = 135 months, 11 years and 4 months = 136 months, and so on.

EXAMPLE 2

Which is a better statistical question for a doctor to ask a patient about exercise? Explain your reasoning.
Do you exercise?   OR   *How many days a week do you exercise?*

SOLUTION

The second question is better because it will give the doctor more specific information. The first question only gives an answer of "yes" or "no". This question does not give any information about whether the patient is getting enough exercise.

When you are collecting data it is important to be aware of **bias** in how the data was collected. Bias is a systematic error that contributes to the inaccuracy of your data.

EXAMPLE 3

Explain how there could be bias in each of the following situations.
  a. A restaurant owner asks customers to rate their dining experience on a scale of 1-10 as they leave the restaurant.
  b. A worker surveys people outside the monkey exhibit at the zoo, asking: "What is your favorite animal at the zoo?"
  c. An ice cream company surveys customers, asking: "What is your favorite type of ice cream: Chocolate, Vanilla or Caramel Fudge Swirl?"
  d. A hotel leaves a comment card in each room where customers can choose to rate their experience at the hotel on a scale of 1-10.

SOLUTIONS

  a. Some customers may not feel comfortable being honest with the restaurant owner.
  b. People outside the monkey exhibit may be more likely to enjoy the monkeys best. Their opinions may not represent all visitors at the zoo.
  c. Only giving customers three options to choose from may not give the most reliable results. Also, the third option sounds more exciting than the other two, so it might be chosen more often.
  d. Optional surveys often receive responses that are really positive or really negative and not many in the middle.

Another way data can be misleading is in how it is displayed. Changing the scale of axes can create very different impressions of a data set.

EXAMPLE 4

Students at Humphrey Middle School collected data on the type of music students would prefer for the next school dance. The bar graph below shows the results.

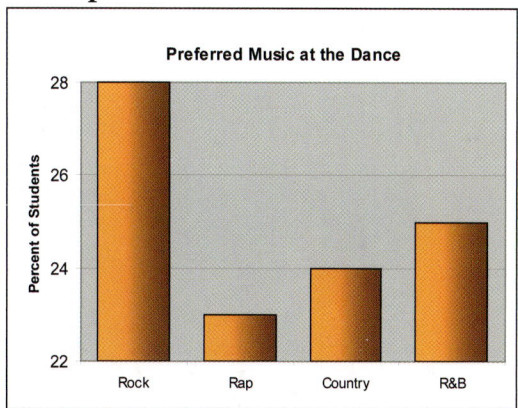

a. What is misleading about the graph?
b. Draw a new bar graph that is not misleading.
c. Why might someone use the original graph to represent the data?

SOLUTIONS

a. The graph does not start at 0%. This makes it appear that Rock had a large majority of the votes. There is really only a five percent difference between Rock and Rap.

b. Starting the graph at 0% would better represent how close the vote was.

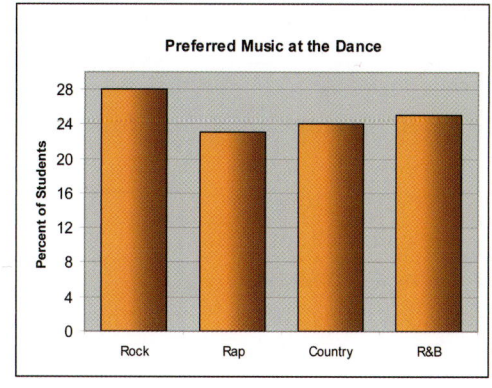

c. Someone who likes Rock music might use this graph. They may want to hear more Rock music at the next school dance.

Lesson 4.1 ~ Introduction to Statistics

EXERCISES

1. What is the difference between numerical data and categorical data?

Classify each data set as either categorical or numerical.

2. the heights of students in your class

3. the favorite food of students in your class

4. the shoe sizes of twelve people

5. the time it takes for students to get to school

6. the maximum speeds of twenty roller coasters

7. number of each type of flower sold by a florist

Determine which is a better statistical question. Explain your reasoning.

8. A. *How many pets do you own?* OR B. *Do you have a pet?*

9. A. *How many points per game does your team score?* OR
   B. *How many games did the school basketball team play this year?*

10. A. *Which teacher has the largest class in your school?* OR
    B. *How many students are in each class?*

11. Which statistical question will give data that is more varied? Explain.
    A. *How many hours do you work per week?* OR B. *How many days do you work per week?*

12. Brad writes for the school newspaper. He surveyed a group of his friends using the questions at the right. He called them late the night before his article was due to the newspaper. Some of his friends did not answer the phone. He drew the following conclusions for his article:
    - Math is the favorite class of students at Happy Rock Middle School.
    - Students have an average of five A's in their classes.
    - Students typically stay up until about 11 pm.

    Students at Happy Rock Middle School
    1. What is your favorite class? (Choose one)
       Science ___ Math ___ Music ___
    2. In how many classes do you have an A?
    3. How late do you stay up on school nights?

    a. Who did Brad survey to get his data? How might this create bias?
    b. Why is the conclusion Brad made from question #1 on his survey not reliable?
    c. Is Brad's data about the number of A's that students are earning accurate? What bias might have been created when Brad asked his friends this question?
    d. Brad made his phone calls late at night. How could this have affected the results of his survey?
    e. If Brad did his survey again, what recommendations would you give him to improve the accuracy of the results?

120   Lesson 4.1 ~ Introduction to Statistics

Describe the possible bias in each survey situation. Explain how to modify each situation to eliminate the bias.

13. An employee asks customers entering a pizza restaurant about their "favorite type of food".

14. The president of a company asks employees if they are "happy with their salary".

15. Respondents answer a telephone survey about whether they plan to vote Republican or Democrat for the next Presidential election.

16. A girl asks her friends how much allowance they receive per month to determine a typical allowance.

17. A hair salon placed a "1-800" phone number on each receipt so customers can call to give feedback on their haircuts.

18. A teacher asks the first one hundred students who arrive at school on Monday what their GPA is in order to determine the typical GPA of students at her school.

19. The line plot at right shows the gas prices at one gas station from 2009 to 2013.
  a. Draw a new line plot which exaggerates the increase in gas prices since 2009.
  b. Which graph (the original or the graph from **part a**) would the station owner use to bring in more customers? Explain your reasoning.
  c. Which graph would a car dealer use to convince customers to buy a hybrid vehicle? Explain your reasoning.

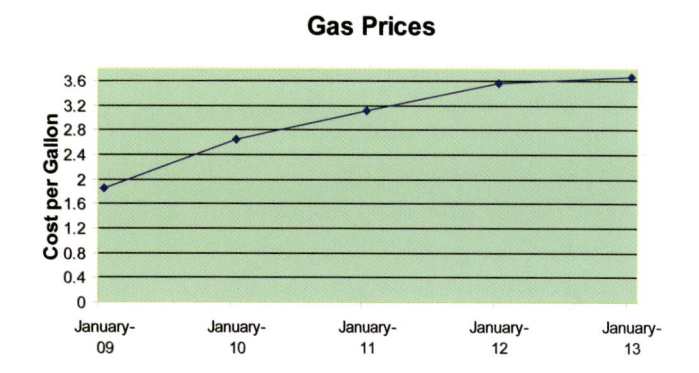

REVIEW

Convert each fraction to a decimal.

20. $\frac{3}{4}$

21. $\frac{2}{5}$

22. $\frac{1}{2}$

23. $\frac{1}{3}$

Round each number to the nearest hundredth.

24. 9.234

25. $0.\overline{3}$

26. 0.4561237

27. $0.2\overline{6}$

Lesson 4.1 ~ Introduction to Statistics

MEASURES OF CENTER

LESSON 4.2

Find measures of center and range.

When looking at data, it is helpful to try to find a single number that can represent all of the values in a data set. There are three numbers which are commonly used to represent a set of numbers. These three numbers are called the **measures of center**.

EXPLORE!

COUNTING PETS

The measures of center can be used to summarize the number of pets that students own. Cameron asked eight of his classmates how many pets they own. The results are listed below.

$$1, 0, 2, 0, 3, 7, 0, 2$$

Step 1: Cameron wants to know the average number of pets that students own. The word "average" refers to the **mean** of the data set. The mean is the sum of all the values divided by the number of values.
  a. Find the sum of all the pets in the list.
  b. Divide the sum by the number of students Cameron asked. This number represents the mean of the data.

Step 2: Another measure of center is called the **median**. When all the numbers have been put in order from least to greatest, the median is the middle number of the ordered data set.
  a. Put the numbers in Cameron's list in order from least to greatest.
  b. Because there are an even number of values in the data set, there will be two middle numbers in the list. Count in from each end to find the two middle numbers. What are they?
  c. When there are two middle numbers, the median is found by finding the mean of the two middle numbers. Add the two middle numbers together and divide by two. This is the median.

Step 3: The last measure of center is the **mode**. The mode is the number(s) or item(s) in a data set which occurs the most often. There can be one mode, no mode or multiple modes. What number of pets showed up most often in Cameron's data set? This is the mode.

Step 4: If Cameron wants to convince his parents that they should get another pet, which measure of center should he use? Explain your choice.

MEASURES OF CENTER

Mean
The sum of all values divided by the number of values

$$\text{Mean} = \frac{\text{Sum of Values}}{\text{Number of Values}}$$

Median
The middle number of an ordered data set

If there are two middle numbers, find the mean of those numbers.

Mode
The number or item in a data set which appears most often

A data set may have one mode, no mode or several modes.

When finding the mean of a set of numbers, use the parentheses on a calculator to group the values before the calculator performs the division operation. For example, to find the mean of 6, 7 and 8 on a calculator:

$$(\; 6 \; + \; 7 \; + \; 8 \;) \; \div \; 3 \; = \; 7$$

EXAMPLE 1 Find the mean, median and mode for the following data sets:
a. 3, 8, 9, 9, 10, 10, 28
b. 17, 8, 10, 15, 19, 2, 9, 12

Total numerator before dividing.

SOLUTIONS

a. Find the mean.
The mean is 11.

$$\frac{3 + 8 + 9 + 9 + 10 + 10 + 28}{7} = \frac{77}{7} = 11$$

Find the median.
3, 8, 9, **9**, 10, 10, 28

The median is 9.

Find the mode.
The modes are 9 and 10.

Both 9 and 10 appear twice.

b. Find the mean.
The mean is 11.5.

$$\frac{17 + 8 + 10 + 15 + 19 + 2 + 9 + 12}{8} = \frac{92}{8} = 11.5$$

Find the median. Order the numbers from least to greatest. Ten and twelve are the middle numbers.

2, 8, 9, 10, 12, 15, 17, 19

Find the mean of these numbers.
The median is 11.

$$\frac{10 + 12}{2} = 11$$

Find the mode.
There is no mode.

No number appears more than any other number.

Lesson 4.2 ~ Measures of Center   123

A useful statistic to describe the spread of a set of data is called the range. The range is the difference between the maximum (largest value) and minimum (smallest value) values in a data set.

Two data sets may have the same mean but different ranges. The data set with the largest range shows that the data is more spread out than the data in the other set.

> **RANGE = MAXIMUM − MINIMUM**
> The maximum is the largest number in the data set.
> The minimum is the smallest number in the data set.

EXAMPLE 2 Find the range of the following data sets.
a. 17, 29, 33, 34, 38, 42
b. 26, 37, 40, 33, 35, 38

SOLUTIONS

a. The maximum is 42 and the minimum is 17.
   Range = Maximum − Minimum
   = 42 − 17
   = 25

b. The maximum is 40 and the minimum is 26.
   Range = Maximum − Minimum
   = 40 − 26
   = 14

EXAMPLE 3 Use the range to find the missing value in each ordered data set.
a. 11, 13, 16, 20, 22, ___   Range = 19
b. ___ , 35, 40, 40, 41, 45, 52   Range = 25

SOLUTIONS

a. The range is 19 and the minimum is 11.

   Add 19 to 11 to find the maximum value.
   The missing value is 30.

   Range = Maximum − Minimum
   19 = Maximum − 11
   19 + 11 = 30

   Think: What number minus 11 is 19?

b. The range is 25 and the maximum is 52.

   Subtract 25 from 52 to find the minimum value.
   The missing value is 27.

   Range = Maximum − Minimum
   25 = 52 − Minimum
   52 − 25 = 27

   Think: 52 minus what number is 25?

Lesson 4.2 ~ Measures of Center

EXERCISES

1. Hector calculated the mean of the following set of numbers: 12, 4, 8, 11, 10. He typed the following into his calculator: 12 + 4 + 8 + 11 + 10 ÷ 5. His result was 37.
   a. Hector knows that his answer is not correct. Describe Hector's error.
   b. Rewrite Hector's expression above with parentheses in the appropriate place.
   c. Find the correct mean for the data set.

2. Chloe played in eight basketball games. She calculated her median to be 15.5 points scored per game. Look at her calculation.
   a. What mistake did Chloe make when finding the median?
   b. Find the correct median for Chloe's points per game.

   Chloe's Work
   9, 20, 17, ⟨17, 14⟩ 20, 19, 6

   $$\frac{17 + 14}{2} = \frac{31}{2} = 15.5$$

3. Kyle and Elena both look at the set of data shown below. Kyle claims that the data set has modes of 12, 16, 17 and 19. Elena claims that the data set has no mode. Who is correct? Explain your reasoning.

   12, 12, 16, 16, 17, 17, 19, 19

Find the mean of each data set.

4. 12, 7, 2, 4, 5

5. 8, 20, 10, 9, 22, 3

6. 10, 3, 3, 7, 5, 9, 12, 3

Find the median of each data set.

7. 4, 9, 13, 14, 17, 18, 18

8. 20, 23, 24, 26, 29, 30

9. 12, 16, 8, 24, 15, 13, 30, 28, 25

Find the mode(s) of each data set. If there is no mode, state "no mode".

10. 42, 45, 48, 52, 52

11. 7, 10, 8, 13, 16, 8, 10, 21

12. 4, 9, 4, 9, 10, 9, 10, 10, 4

Find the range of each data set.

13. 2, 5, 6, 9, 12, 15

14. 25, 38, 32, 40, 29, 39

15. 14, 26, 39, 55, 12, 40, 32

Find the three measures of center. Round answers to the nearest tenth, if necessary.

16. 18, 19, 20, 28, 30

17. 5, 5, 8, 10, 10, 25

18. 7, 8, 9, 10, 11, 12, 13

19. 10, 12, 20, 22, 22, 12, 20

20. 18, 15, 16, 25, 21, 20, 15, 22, 19

21. 6, 10, 14, 11, 12, 8, 5, 7

Lesson 4.2 ~ Measures of Center   125

Use the range to find the missing value in each ordered data set.

22. 13, 21, 26, 26, ___   Range = 18   **23.** ___, 32, 38, 41, 45, 51   Range = 24

24. ___, 17, 22, 29, 33, 36, 42   Range = 32   **25.** 55, 62, 68, 77, 81, 84, ___   Range = 43

26. Naomi wants to earn an A (90%) in her math class. On her first three tests, she scored 87%, 98% and 86%. What score will she need to earn on her fourth test in order to have an average of 90%? Show all work necessary to justify your answer.

27. Braden earns money on weekends by mowing lawns. The last four weekends he has earned $20, $24, $22 and $24. How much will he need to earn next weekend to average $25 per weekend? Use mathematics to justify your answer.

Use the range and the mean to determine the missing numbers in each ordered data set. Show all work necessary to justify your answers.

28. 6, ___, 12, 14, ___
Mean = 13
Range = 15

29. ___, 24, 33, 35, ___, 52
Mean = 36
Range = 32

30. Create a data set with four numbers that has a median of 10 and a mean of 12.

31. Kiesha saw in the paper that the median age in her city is 28 and the mean age is 35. What generalizations can you make about the population of her city based on these two statistics?

REVIEW

Convert each fraction to a decimal.

32. $\frac{3}{5}$ **33.** $\frac{2}{3}$ **34.** $\frac{1}{4}$

35. Which of the following is a better statistical question? Explain your reasoning.
     A. *Are fruits nutritious?* OR
     B. *How many calories are in each fruit?*

36. Write a statistical question that would give categorical data.

37. Write a statistical question that would give numerical data.

38. Describe the possible bias in the following survey situation:
Calling people in the middle of the day to collect data about how many hours they work per week.

Tic-Tac-Toe ~ Shifting Data

Answer the questions below to see how changing the values in a data set affect the mean and median.

Katrina enjoys figure skating. She goes to a lot of competitions around the state. At her latest competition, she performed her routine and received the following scores from the eight judges:

7.5, 7.5, 7.5, 8.0, 8.0, 8.5, 9.0, 9.0

1. Find the mean, median and mode for Katrina's scores.

2. After all the other skaters completed their routines, the announcer notified everyone that there had been an error in the judges' scoring. Each skater actually needed to add 1.0 to each of their scores to get their new score. What are Katrina's new scores?

3. What is Katrina's new mean, median and mode?

4. How do the answers in #3 compare to those in #1? In general, how does adding to or subtracting from each number in a data set affect the mean and median?

Katrina and her friend, Carmen, have entered another skating event. Here were their scores for the competition:

Katrina: 7.0, 7.5, 8.0, 8.0, 8.5, 8.5, 8.5, 9.0
Carmen: 7.0, 7.0, 8.0, 8.0, 9.0, 9.0, 9.5, 10.0

5. Find the mean and median for Katrina's scores.

6. Find the mean and median for Carmen's scores.

Under new contest rules, the high and low score for each skater is dropped. For example, Katrina's scores of 9.0 (high) and 7.0 (low) would get dropped.

7. What will Katrina's new mean and median be after her high and low scores get dropped?

8. How does Katrina's mean and median in #7 compare to her statistics in #5?

9. Which of Carmen's scores will get dropped under the new rules?

10. Find Carmen's new mean and median after the scores are dropped.

11. How does Carmen's mean and median in #10 compare to her statistics in #6?

12. Summarize your findings from #8 and #11. Which measure of center is not affected by dropping the high and low data values? Why do you think this is so?

Lesson 4.2 ~ Measures of Center

DOT PLOTS

LESSON 4.3

Display data using dot plots.

A **dot plot** is a useful way to visually display the spread of data. It consists of a number line with dots equally spaced above each value. The number of dots represents the number of times each value is in the data set. In the dot plot below there are three dots above the number 6. This means there are three people that have six fish. Clusters and gaps in the data are easy to see in a dot plot.

Number of Fish in Students' Aquariums

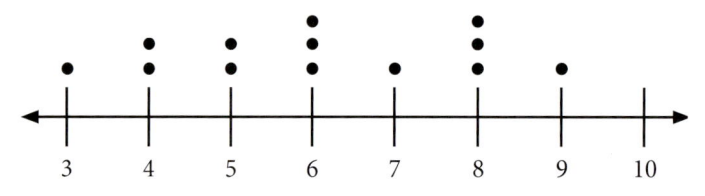

EXAMPLE 1

The dot plot below shows the number of siblings for each student in Darla's class. Use the plot to answer the questions below.

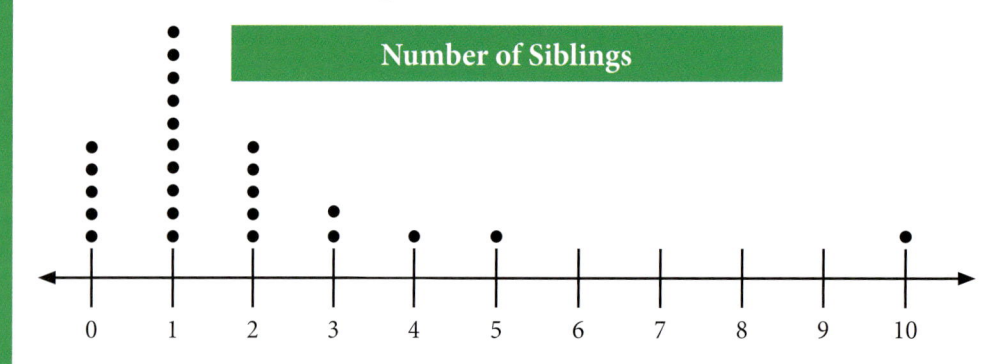

Number of Siblings

a. How many students are in Darla's class?
b. What is the mode of the number of siblings?
c. How many students have either 0 siblings or 1 sibling?
d. What percent of students have 3 or more siblings?

SOLUTIONS

a. Count the number of dots in the dot plot. There are 25 students in Darla's class.
b. Look for the largest stack of dots in the dot plot. There are ten people with 1 sibling. One sibling is the mode of the data set.
c. Count the number of dots above the 0 and 1. There are 15 students with either 1 sibling or no siblings.
d. There are 5 (2 + 1 + 1 + 1) students with 3 or more siblings and 25 total students in the class.

$$\frac{5}{25} = 5 \div 25 = 0.2 \quad \rightarrow \quad 0.2 \cdot 100 = 20\%$$

20% of students have 3 or more siblings.

EXPLORE!   **HOW TALL?**

Step 1: Find your height to the nearest inch and record your answer using only inches. If you are 5 feet 2 inches tall, for example, you would use the process below to convert your measurement to inches:

        5 feet · 12 inches per foot = 60 inches
        Add the 2 extra inches (60 + 2)
        5' 2" = 62 inches tall

Step 2: Record the height data for all the students in your class, including your own.

Step 3: What are the minimum (lowest) and maximum (highest) values in the height data? What is the range of the data?

Step 4: Make a number line that extends from the minimum to the maximum value in **Step 3**. Evenly space "tick marks" along the number line for each of the values between the minimum and maximum.

Step 5: Using your data set from **Step 2**, put a dot above the number line for each data value. If a data value occurs more than once, stack the dots as shown in **Example 1**.

Step 6: Describe the spread of your data. Are your values clustered together or are they spread out?

Step 7: Based on your dot plot, what is the mode of the data? How can you tell by looking at your dot plot?

Step 8: What is the median of your data set? How does the median compare with the mode?

Michelle made three dot plots to show the answers to her statistical question. Which dot plot is drawn correctly?

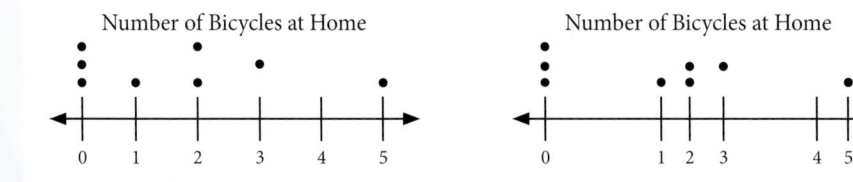

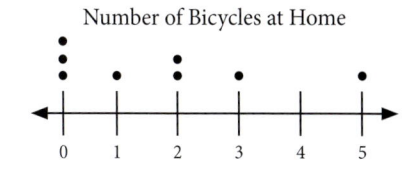

The first dot plot looks as if there are as many people with 2 bicycles as those with 0 bicycles. This is because the dots are not equally spaced above the number line. The second dot plot does not have the values equally spaced along the number line. The third dot plot is drawn correctly.

Lesson 4.3 ~ Dot Plots   **129**

EXAMPLE 2

Sydney asked ten of her friends the number of states that they have visited.
These were their answers: 3, 0, 1, 4, 2, 0, 2, 1, 7, 2

a. Make a dot plot of the results.
b. What are the mean, median and mode of the data?

SOLUTIONS

a. The minimum value is 0 and the maximum is 7. Create a number line that extends to these values. Place dots above each value, stacking dots when a value occurs more than once.

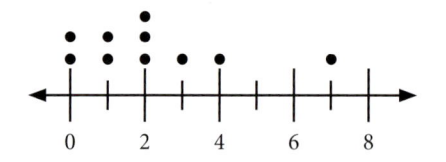

b. To find the mean, divide the sum of all answers by the total number of answers.

$$\frac{3+0+1+4+2+0+2+1+7+2}{10} = \frac{22}{10} = 2.2$$

The mean is 2.2.

To find the median list the values from least to greatest.

0 0 1 1 2 2 2 3 4 7

Find the average of the 5th and 6th values.

$$\frac{2+2}{2} = 2$$

The median is 2.

The mode is the value which shows up the most often.

The mode is 2.

EXERCISES

1. Make a dot plot to represent each data set:
   a. 10, 5, 7, 2, 5, 8, 8, 9, 7, 1, 9, 10, 8, 6, 8, 9
   b. 62, 66, 68, 76, 68, 70, 68, 80, 68, 70, 66, 64, 68, 66, 70

2. Why is it important to equally space the dots above each value in a dot plot?

3. Taylor recorded the pulse rates (in beats per minute) of 12 of his classmates:
   69, 92, 64, 78, 80, 82, 86, 93, 75, 66, 74, 71
   a. Make a dot plot of the data Taylor collected.
   b. Do you think a dot plot is a good way to display this data? Explain your reasoning.

4. Explain whether or not each data set below should be displayed in a dot plot.
   a. The number of cell phones in students' households.
   b. The cost of 20 new cars in a car lot.
   c. The year students in your class were born.

5. The table shows the number of times students in Mrs. Arrington's class went to the mall in the last month.
   a. Construct a dot plot to display the number of times that students went to the mall.
   b. What is the median value for the class?
   c. How many students are in the class?
   d. How many total visits to the mall did students in Mrs. Arrington's class make in the last month?
   e. What is the average (mean) number of mall visits for Mrs. Arrington's students?

Number of Visits to the Mall	Number of Students
0	5
1	8
2	4
4	3
5	1
6	1

6. Which graph below would best represent each data description? Explain how you know your answer is correct.
   a. Number of times students in the class have flown.
   b. Speed (in miles per hour) of cars traveling down a highway.
   c. Number of people living in students' homes.
   d. Percent that students earned on the last math test.

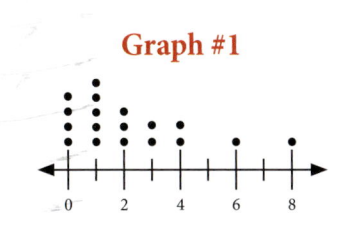

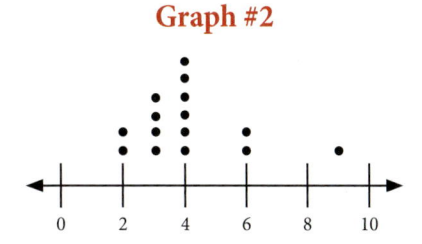

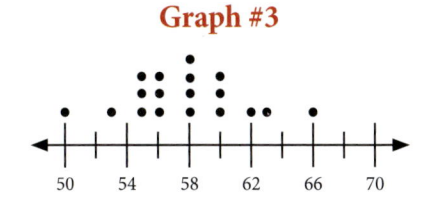

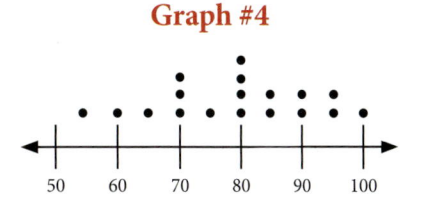

Graph #1

Graph #2

Graph #3

Graph #4

7. Draw a dot plot that could represent the heights, in inches, of twelve players on a basketball team. Draw another dot plot that could represent the heights of twelve random adult men. How do the dot plots compare?

8. Draw a dot plot that shows ten values having a mean of 5, a median of 5 and a range of 10.

9. Carmen says it is easy to find the median of a data set on a dot plot. All you have to do is find the column with the most dots! Carmen is not correct.
   a. Identify which measure of center Carmen described finding in a dot plot.
   b. Explain how you find the median of the data in a dot plot.

Lesson 4.3 ~ Dot Plots

10. The dot plot below shows the length, in inches, of fifteen babies born today at a hospital.

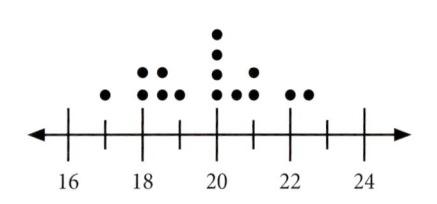

Length of Baby (inches)

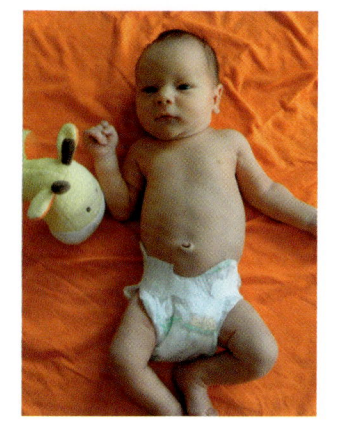

  a. What is the range of the data?
  b. What length appears most often?
  c. What is the median baby length?
  d. What is the mean baby length?
  e. What percent of babies born were at least 20 inches in length?

11. A police officer noted the following about the speed (in miles per hour) of twenty cars on a stretch of road on the interstate. Draw a dot plot that includes each description below.

- half of the cars drove at least 65 mph
- 20% of the cars drove over 70 mph
- 30% of the cars drove 60 mph or slower
- the most common speed traveled was 60 mph

12. Cameron surveyed ten of his classmates. He asked them how much pocket change they had and he noted the following characteristics about the data. Draw a dot plot that includes each description at the right.

- the mode and minimum was $0
- the range was $0.80
- half the students had $0.15 or less
- the mean was $0.26

13. Ask ten people how many siblings they have. Record your data in a dot plot. Find the mean, median, mode and range of your data.

REVIEW

Find the three measures of center. Round answers to the nearest tenth, if necessary.

14. 10, 13, 14, 15, 20, 22, 25

15. 20, 18, 32, 43, 20, 7, 28, 32

Use the range to find the missing value in each ordered data set.

16. ___, 16, 22, 23, 28, 32, 35
Range = 24

17. 45, 62, 64, 72, 75, 77, ___
Range = 34

Classify each data set as either categorical or numerical.

18. the time that students wake up in the morning on a school day

19. type of transportation for students to get to school

20. Which statistical question will give more varied answers? Explain your reasoning.
   A. *How many yards can you throw a football?*   OR   B. *How many feet can you throw a football?*

Tic-Tac-Toe ~ Make It Complete!

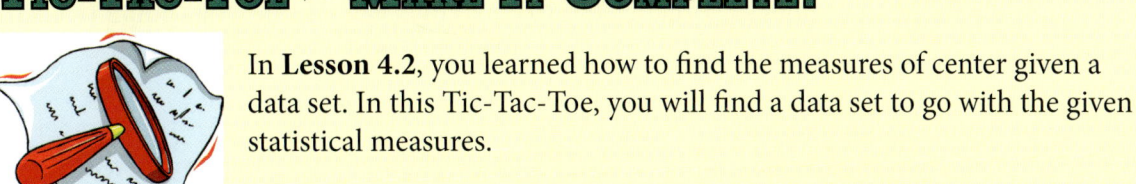

In **Lesson 4.2**, you learned how to find the measures of center given a data set. In this Tic-Tac-Toe, you will find a data set to go with the given statistical measures.

1. Use the partial data set and the statistics provided to find the remaining data pieces: 25, 33, 20, ____, ____, ____
   Mean = 27   Median = 26   Mode = None   Range = 14

2. Justin's teacher gave him the following statistics about his six test scores:
   Mean = 84   Median = 84   Mode = 86   Range = 26
He recalls that his lowest test score was a 72. What were his six test scores?

Create a data set with five numbers and the following statistics. There are many possible answers.

3. Mean = 8 and Range = 10

4. Mean = 12 and Range = 15

Note: Exercises 5 and 6 can have multiple answers.

5. Jack planted seven seeds for bean plants. He let the plants grow for 2 months and then collected the following statistics about his plants. The mean height of his plants was 11 inches. The median height was 8 inches and the mode was 15 inches. The smallest plant grew to only 2 inches and the range of their heights was 26 inches. Give one possible data set for the heights of Jack's seven bean plants.

6. Tamara has to monitor the weight of her five turkeys for her 4-H project. The turkeys have a mean weight of 12.5 pounds, a median of 13 pounds and a maximum of 15.2 pounds. The range of their weights is 6.8 pounds. What are the possible weights for her turkeys?

Tic-Tac-Toe ~ New Student Brochure

A new student has just joined the class. Make a brochure describing how to find the mean, median and mode of data sets. Include the definitions of each as well as examples showing how to find the values. Use pictures, symbols and markings to illustrate the concepts. Also, include a section on how to find these values from dot plots.

On a separate sheet of paper, create a short quiz (6-10 questions) that the student could take to demonstrate knowledge of these concepts. Create a key for the quiz as well.

HISTOGRAMS

LESSON 4.4

Make, read and interpret histograms.

When the numbers in a data set are more spread out and there are not a lot of repeating values, a dot plot may not be the best way to display the data. A **histogram** is another helpful way to display a data set. A histogram is a bar graph in which data values are organized into equal intervals. While a bar graph typically displays data in categories, a histogram displays data in the form of numbers.

Ms. Sanchez collected data about students and their pets. The first graph, a bar graph, shows the number of each type of pet her students have. The second graph is a histogram. What do you see that is different?

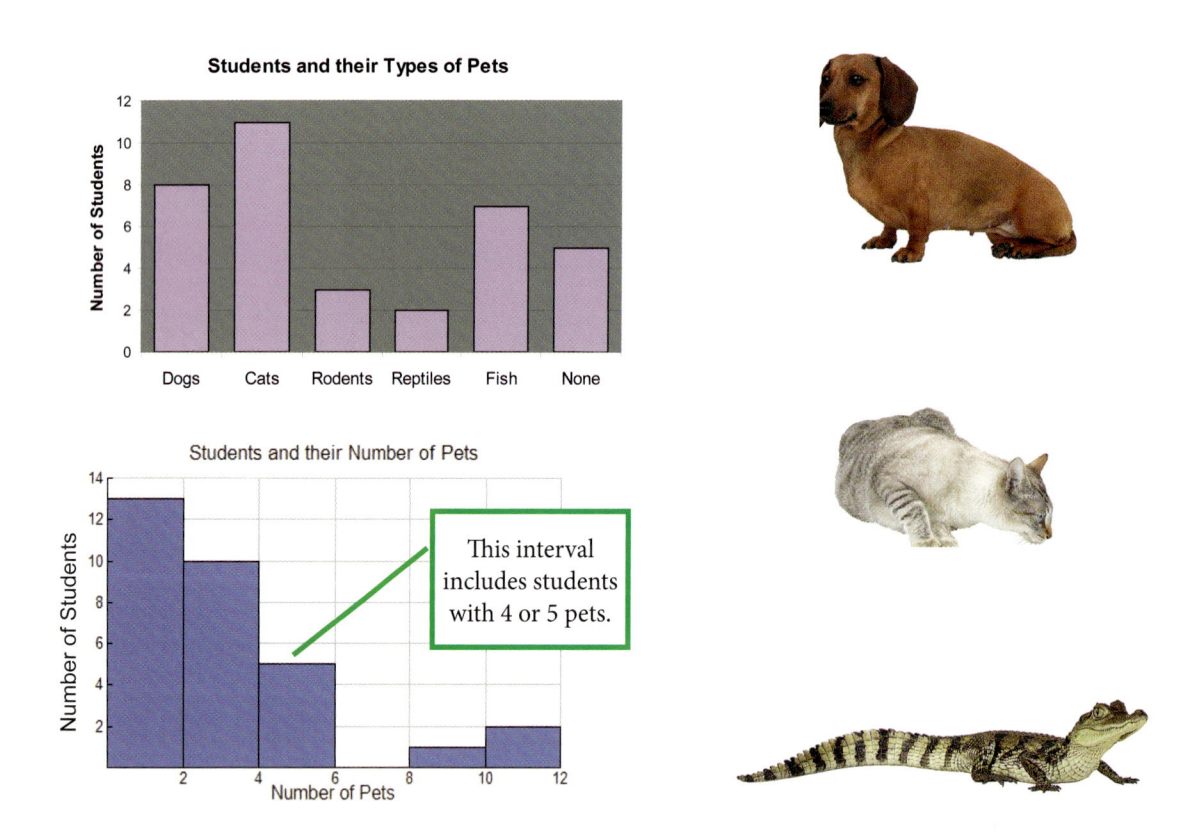

This interval includes students with 4 or 5 pets.

The histogram focuses on the number of pets that each student has rather than the type of pets. There are two key parts of the histogram: the intervals along the horizontal axis of the graph and the number of students that fall into each interval along the vertical axis.

Each interval on a histogram includes the number on the left-hand side of the interval up to the number on the right-hand side. For example, the first bar in the histogram includes all students that have 0 or 1 pet. Students who have 2 pets are included in the second interval.

134   Lesson 4.4 ~ Histograms

EXAMPLE 1

Use the histogram to answer each question.
a. How many students had 2 or 3 pets?
b. How many students had at least 8 pets?
c. The interval between 6-8 is empty. What does that mean?
d. How many students are in Ms. Sanchez' class?

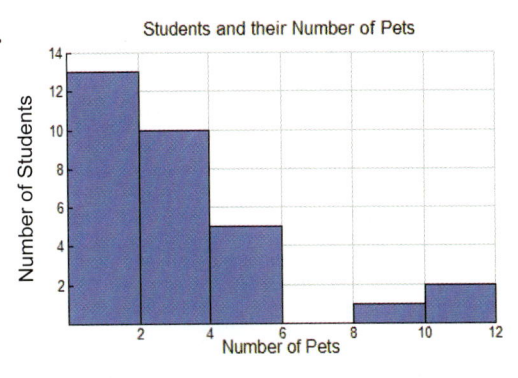

SOLUTIONS

a. Ten students have 2 or 3 pets. Look at the height (number of students) of the bar that is between 2 and 4.

b. There are 3 students who have at least 8 pets. Adding the bars to the right of 8 on the horizontal axis (1 + 2) totals 3.

c. This means that there are 0 students who had exactly 6 or 7 pets.

d. Add the heights of all the bars together. This sum represents the number of students in the class.
$$13 + 10 + 5 + 0 + 1 + 2 = 31 \text{ students}$$

Before making the histogram, Ms. Sanchez made a **frequency table**. A frequency table shows a tally of how many times a value occurred in each interval. Ms. Sanchez tallied the students who had the number of pets in the intervals.

Number of Pets	Tally
0-2	⋕⋕ ⋕⋕ III
2-4	⋕⋕ ⋕⋕
4-6	⋕⋕
6-8	
8-10	I
10-12	II

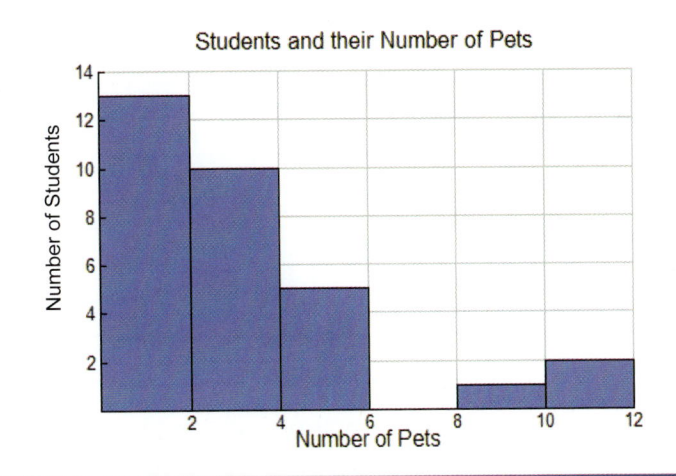

MAKING A HISTOGRAM

1. Identify the minimum and maximum numbers in the data set.
2. Choose an appropriate interval width. Histograms usually have 4 to 12 intervals.
3. Make a frequency table. Tally the number of times a value occurs in each interval.
4. Place the intervals from the frequency table along the horizontal axis of the graph.
5. Create bars for each interval. The height of each bar should equal the number of tally marks in each interval in the frequency table.

Lesson 4.4 ~ Histograms

Any interval width works for histograms. However, it is ideal to have between 4 and 12 intervals to best display the distribution of data in a histogram. Too few intervals in a histogram will not show the distribution in the graph well. Too many bars will spread the data out so the bars will have very little height.

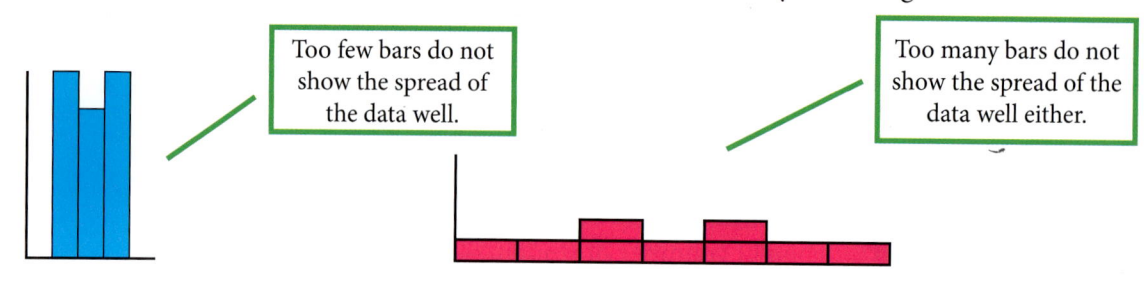

Too few bars do not show the spread of the data well.

Too many bars do not show the spread of the data well either.

EXAMPLE 2

Make a histogram for students' pulse rates (in beats per minute) listed below.
64, 76, 60, 70, 68, 70, 88, 62, 54, 68, 70, 72, 60, 92, 64, 76, 70, 68, 62, 72

SOLUTION

Identify the minimum and maximum values.

Minimum = 54
Maximum = 92

The frequency table must start low enough to include 54 and must go high enough to include 92.

The overall range of the data is 92 − 54 or 38.
Determine an appropriate interval width. $38 \div 8 \approx 5$
An interval width of 5 will create about 8 bars.

Complete the frequency table for the data.

Pulse Rates	Number of Students Tally
50 - 55		
55 - 60	
60 - 65	|||||	
65 - 70	|||
70 - 75	|||||
75 - 80	||
80 - 85	
85 - 90		
90 - 95		

When a number in the data set lies on the border of two intervals, put it in the uppermost interval.

Use the information collected in the frequency table to draw the histogram.

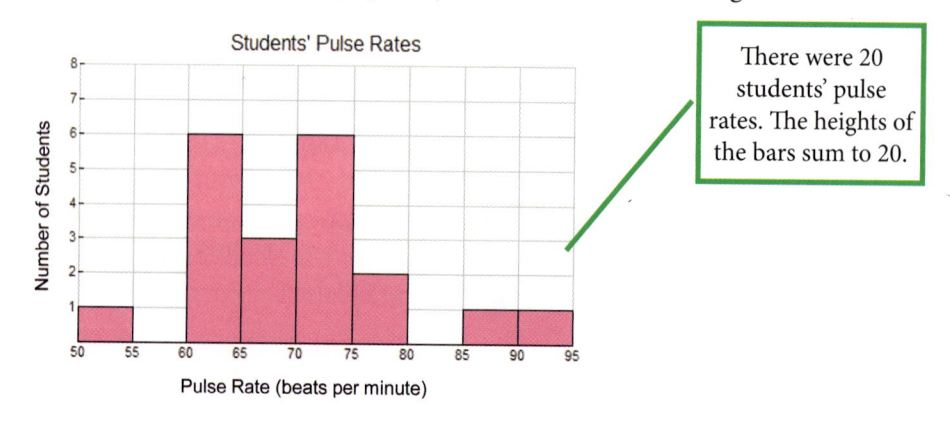

Students' Pulse Rates

There were 20 students' pulse rates. The heights of the bars sum to 20.

Pulse Rate (beats per minute)

136   Lesson 4.4 ~ Histograms

EXERCISES

1. Explain the difference between a bar graph and a histogram.

2. The number of miles driven on seventeen cars was recorded. The data was displayed in the histogram below. Use the graph to answer the questions.

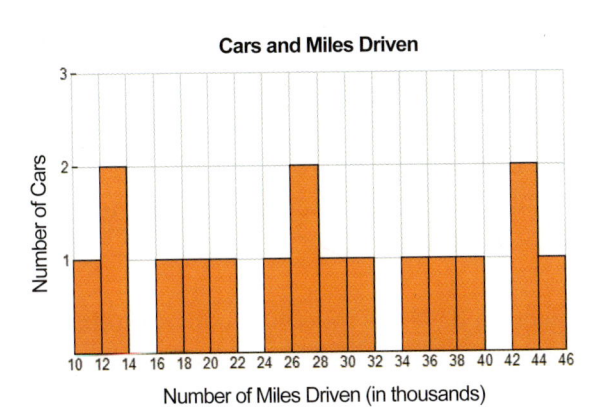

   Cars and Miles Driven

   a. How many intervals are shown in the histogram (be sure to include the empty intervals in the total)?
   b. Is the interval width appropriate? Explain your reasoning.
   c. Copy the frequency table below. Use the information in the histogram above to complete the table.

Number of Miles Driven (in thousands)	Number of Cars
10 – 20	
20 – 30	
30 – 40	
40 – 50	

   d. Draw a new histogram with the interval width used in **part c**.

3. Newborn babies were weighed and the data was recorded in the frequency table below. Use the table to answer the following questions.

Weight of Newborn (in pounds)	Tally
4 – 5.5	|
5.5 – 7	|||
7 – 8.5	|||||
8.5 – 10	||
10 – 11.5	|

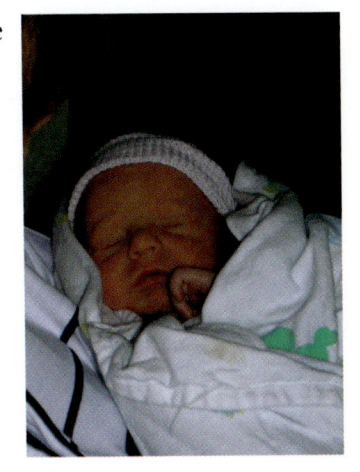

   a. How many babies have been tallied so far?
   b. What is the interval width?
   c. Glen thought a 7-pound baby should be tallied in the 5.5 – 7 interval. Eliza claimed it should be tallied in the 7 – 8.5 interval. Who is correct? Explain your reasoning.
   d. A 10-pound baby boy is born. In which interval should he be tallied?

Lesson 4.4 ~ Histograms   137

4. Ben collected and weighed 20 frogs for his science class. He needs to do a report showing the weight distribution of frogs. Here is the data he collected. Follow the steps below to create a histogram for the data.

Weights of Frogs (in ounces)

5.2	5.9	6.7	4.0	6.3	7.0	6.7	7.2	7.9	5.8
6.0	6.7	7.2	6.8	6.2	6.5	7.1	7.4	6.2	6.6

  a. Find the minimum and maximum values in the data set. What is a reasonable interval width to use for this data set?
  b. Use your interval width in **part a** to create a frequency table for the data. Be sure that each data value is included in only one interval.
  c. Use your frequency table to create a histogram. Be sure to label both axes and title the frequency table.

5. Dierdre collects old movies. The fifteen movies in her collection have the following copyright dates. Choose an appropriate interval width. Make a frequency table and a histogram using Dierdre's movie data.

  1958, 1963, 1945, 1966, 1974, 1958, 1962, 1968,
  1950, 1963, 1959, 1969, 1948, 1965, 1962

6. Stephanie buys purses. She paid the following prices for twelve different purses. Choose an appropriate interval width. Make a frequency table and histogram for Stephanie's purse data.

  $17, $11, $22, $10, $13, $10,
  $27, $14, $49, $12, $18, $20

7. The histogram below shows the number of hours students in Mr. Underhill's class slept last night. Use the histogram to answer each question.
  a. How many students were included in the survey?
  b. What is the interval width?
  c. How many students slept less than 4 hours last night?
  d. Which interval included the most students?
  e. How many students slept between 4 and 8 hours last night?

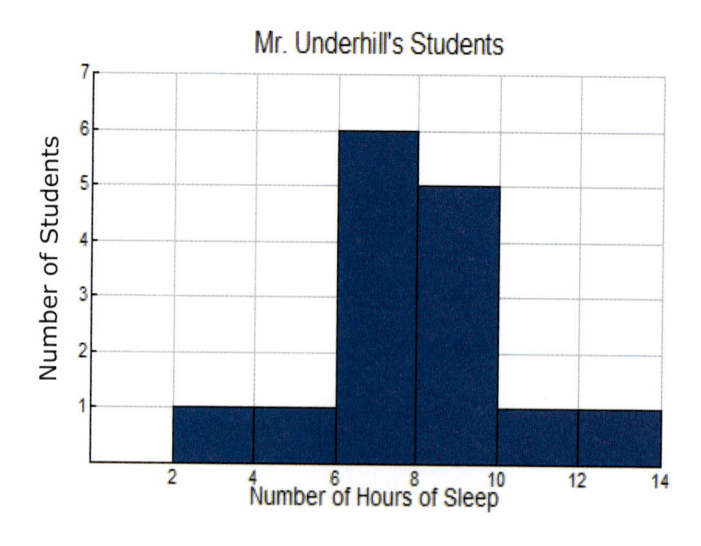

8. The ages of the fourteen people at the local coffee shop on Saturday morning are listed below.

15, 22, 37, 40, 8, 14, 21,
25, 32, 23, 46, 26, 30, 28

   a. What are the minimum and maximum values in the data set?
   b. Create a histogram with an interval width of 5.
   c. Create a histogram with an interval width of 10.
   d. Which histogram would give the owner of the coffee shop the most information about his customers? Explain your reasoning.

9. The histogram below shows the number of visitors to the Children's Museum in one hour.

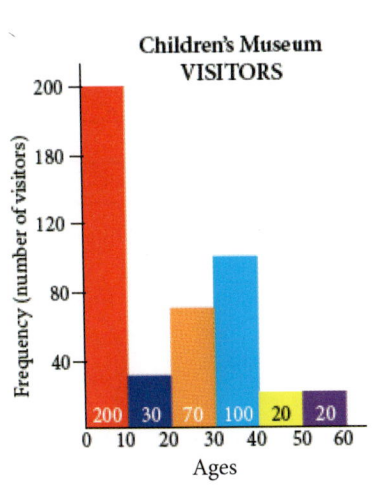

   a. Which age range represented the largest group?
   b. Approximately how many total visitors came to the Children's Museum during that hour?
   c. What percent of the visitors were less than 20 years old? Show all work necessary to justify your answer.

10. A company tracked the number of songs downloaded daily from their music website. Data from the last ten days is shown in the table at right.
    a. Draw a histogram for the data. Let the intervals start at 0 and increase by 1,000.
    b. In what interval did you include the Day 7 data of 3,000 downloads? Why?
    c. Based on your histogram, what range of downloads was the most common?
    d. What percent of days had more than 2,000 songs downloaded?

Day	Songs Downloaded
1	1,294
2	3,810
3	1,984
4	2,458
5	1,723
6	2,917
7	3,000
8	982
9	1,450
10	2,725

11. Peter surveyed a random group of 50 students at his middle school and asked them their age in years.
    a. Why would a histogram NOT be the best data display for him to use?
    b. What is a survey question he could have used that would most likely provide data that could be displayed nicely in a histogram?

Lesson 4.4 ~ Histograms

12. Tia says any data that can be displayed in a dot plot can be shown in a histogram. Is she correct? Explain your reasoning.

13. Lance says he can calculate the mean, median and mode from data displayed in a histogram. Is he correct? Explain your reasoning.

REVIEW

14. The table shows the number of students in history class who have been absent each amount of days.
   a. Construct a dot plot to display the data. Your number line should show the number of absences.
   b. What is the median value for the class?
   c. How many students are in the class?
   d. How many total absences have there been?
   e. What is the average (mean) number of absences for students in this history class?

Number of Absences	Number of Students
0	3
1	9
2	8
3	3
5	2
8	1

Find the range of each data set.

15. 19, 22, 26, 29, 33, 38, 42

16. 60, 72, 58, 65, 76, 66, 70, 65

17. Polly has run the 400-meter race in the following times this year:
   Time (seconds): 52, 54, 56, 53
She wants to have an average time of 53 seconds. What time does she need to get on her fifth race to have a 53-second average? Show all work necessary to justify your answer.

Tic-Tac-Toe ~ Ex"skew"se My Leaning

Research the concept of "skew" as it relates to data displays. What does it mean for a data set to be "skewed right" or "skewed left" or to have a "normal distribution"?

Create a poster that shows examples of histograms and box-and-whisker plots with each type of skew (six examples altogether). Give an example of a real-world data set that would have these types of data displays.

Lesson 4.4 ~ Histograms

BOX-AND-WHISKER PLOTS

LESSON 4.5

Make, read and interpret box-and-whisker plots.

Graphs provide a great way to visualize the spread of the values in a data set. They allow you to analyze the information in more detail than just simply stating the mean, median and mode. For example, the following data shows the number of cell phone minutes used each week by two students:

Kevin: 71, 74, 74, 81
Marcie: 54, 74, 74, 98

Look at the measures of center for both students. Both students had a mean of 75, a median of 74 and a mode of 74. That does not tell the whole story about both students' cell phone usage. What differences do you see in the data sets? Which student can you more accurately predict future minutes used in a week?

Measures of center will not always be the best way to describe a set of data. Other measures and graphs can be used to analyze data. The **five-number summary**, for example, describes the spread of numbers in a data set.

FIVE-NUMBER SUMMARY

Minimum ~ 1st Quartile (Q1) ~ Median ~ 3rd Quartile (Q3) ~ Maximum

---

Kevin kept track of the total minutes he used each month for seven months. Find the five-number summary of the minutes he used by following the steps below.

Kevin: 300, 284, 280, 310, 300, 270, 295

1. Put the numbers in order and find the median.	270, 280, 284, **295**, 300, 300, 310 Median
2. Find the median of the lower half of the data (this is called the **1st Quartile**).	270, **280**, 284, **295**, 300, 300, 310 Q1 Median
3. Find the median of the upper half of the data (this is called the **3rd Quartile**).	270, **280**, 284, **295**, 300, **300**, 310 Q1 Median Q3
4. Identify the minimum and maximum values.	**270**, **280**, 284, **295**, 300, **300**, **310** Min Q1 Median Q3 Max

Do not include the median in the upper or lower half.

The five-number summary of Kevin's cell phone usage data is: 270 ~ 280 ~ 295 ~ 300 ~ 310.

The word "quartile" refers to how the data is broken up into quarters. In Kevin's data, 25% of the time he used between 270 and 280 minutes, 25% he used between 280 and 295 minutes, and so on.

270   280   295   300   310

25%   25%   25%   25%

EXAMPLE 1 Find the five-number summary of the following data sets.
a. 12, 14, 19, 20, 24, 24, 28, 30
b. 10, 9, 9, 8, 6, 5, 9, 10, 8

SOLUTIONS

a. Find the median of the data set.

$$\frac{20+24}{2} = \frac{44}{2} = 22$$

12, 14, 19, 20, | 24, 24, 28, 30
         22
       Median

Find the 1st quartile. If there are two numbers in the middle, include one in each half of the data.

$$\frac{14+19}{2} = \frac{33}{2} = 16.5$$

12, 14, | 19, 20, | 24, 24, 28, 30
       16.5
        Q1

Find the 3rd quartile.

$$\frac{24+28}{2} = \frac{52}{2} = 26$$

12, 14, | 19, 20, | 24, 24, | 28, 30
                          26
                           Q3

Find the minimum and maximum.

12, 14, | 19, 20, | 24, 24, | 28, **30**
Min   Max

The five-number summary is 12 ~ 16.5 ~ 22 ~ 26 ~ 30.

b. Put the numbers in order.

5, 6, 8, 8, 9, 9, 9, 10, 10

Find the median.

5, 6, 8, 8, **9**, 9, 9, 10, 10
           Median

Find the 1st quartile. When there is an odd number of values in the set, do not include the median in either half.

5, 6, | 8, 8, **9**, 9, 9, 10, 10
     7
     Q1

Find the 3rd quartile.

5, 6, | 8, 8, **9**, 9, 9, | 10, 10
                         9.5
                          Q3

Find the minimum and maximum.

5, 6, | 8, 8, **9**, 9, 9, | 10, **10**
Min   Max

The five-number summary is 5 ~ 7 ~ 9 ~ 9.5 ~ 10.

142 Lesson 4.5 ~ Box-and-Whisker Plots

In a previous lesson you learned how to use the range to describe the spread of a data set. Another similar statistic which gives a measure of spread is the **interquartile range** or **IQR**. The IQR is the range of the middle half of the data.

> **INTERQUARTILE RANGE**
>
> The interquartile range (IQR) is the difference between the third quartile and the first quartile in a set of data.
>
> IQR = Q3 − Q1

EXAMPLE 2

Sonny plays basketball. The points he scored in his past eleven games are listed below.

6, 10, 11, 11, 12, 14, 15, 15, 15, 20, 23

a. Find the five-number summary.
b. Find the range and interquartile range of Sonny's points scored.

SOLUTIONS

a. The middle number of the data set (the median) is 14. The median of the lower half of the data (Q1) is 11. The median of the upper half (Q3) is 15. The minimum is 6 and the maximum is 23.

6, 10, 11, 11, 12, 14, 15, 15, 15, 20, 23
Min   Q1   Median   Q3   Max

The five-number summary is 6 ~ 11 ~ 14 ~ 15 ~ 23.

b. Find the range of the data. Range = Maximum − Minimum = 23 − 6 = 17

Find the interquartile range. IQR = Q3 − Q1 = 15 − 11 = 4

A **box-and-whisker plot** is used to display a five-number summary. A box-and-whisker plot visually displays the spread of the data. It shows when groups of numbers are clustered together as well as when the numbers are spaced apart.

The five-number summary divides a data set into four quartiles. Each section represents 25% of the data. A box-and-whisker plot for a set of data makes it easier to answer questions about the distribution of the data.

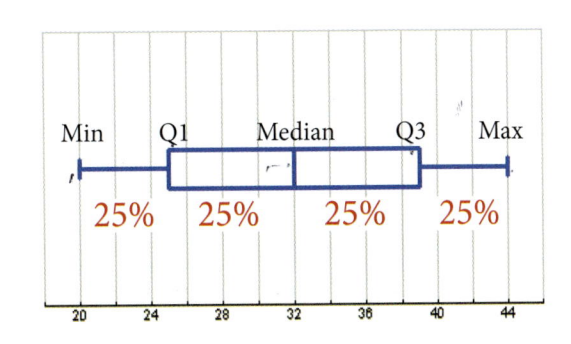

Lesson 4.5 ~ Box-and-Whisker Plots

EXAMPLE 3

A grocery store manager collected data about the amount of money spent by the first eleven customers on a given day.

Amount Spent: $1, $4, $5, $9, $10, $10, $16, $20, $25, $48, $68

a. Construct a box-and-whisker plot to display the amounts spent by the customers.
b. Complete the following statement: "Fifty percent of the customers spent between $5 and $____."

SOLUTIONS

a. Find the five-number summary of the data.

1, 4, **5**, 9, 10, **10**, 16, 20, **25**, 48, **68**

1 ~ 5 ~ 10 ~ 25 ~ 68

Draw a number line. Create equal intervals on your number line that include the minimum (1) and maximum (68) data values. For this data set, a number line spanning from 0 to 70 with intervals of 5 works well.

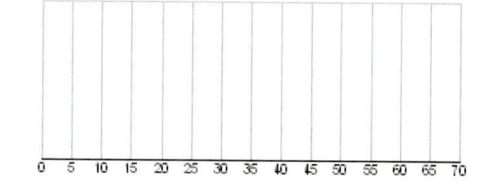

Create a box just above the number line that goes from the Q1 value (5) to the Q3 value (25). Draw a vertical line through the box where the median value (10) lies.

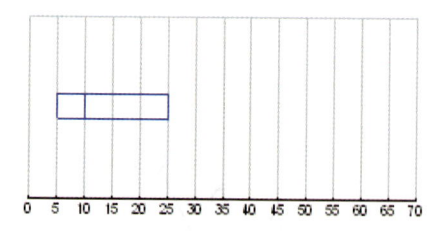

Add "whiskers" to the ends of the box that extend out to the minimum and maximum values.

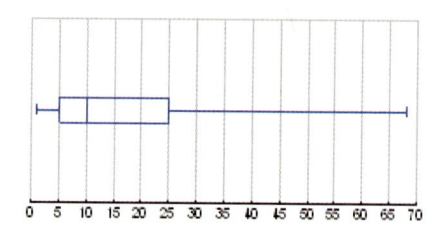

b. The sections of the box and the whiskers each represent 25% "quartiles".
"Fifty percent of the customers spent between $5 and $25."

Lesson 4.5 ~ Box-and-Whisker Plots

EXERCISES

1. List the names of the five numbers used to make a box-and-whisker plot.

2. What does the vertical line in the middle of the box in a box-and-whisker plot represent?

3. Callyn found the five-number summary of a data set. It was 7 ~ 9 ~ 15 ~ 16 ~ 21. She made the box-and-whisker plot shown at the right.
   a. What mistake did Callyn make in her box-and-whisker plot?
   b. Draw a correct box-and-whisker plot for the five-number summary.

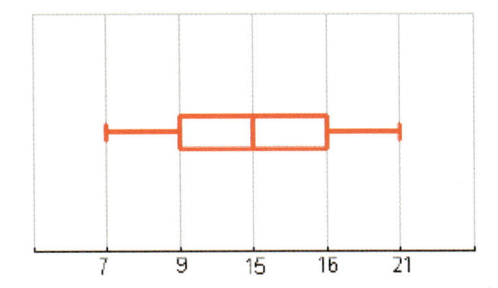

4. Misty found the five-number summary for the following set of data.
   Number of students in each math class: 19, 23, 24, 26, 27, 29, 30
   19 ~ 23.5 ~ 26 ~ 28 ~ 30
   a. Antwan tells Misty her answer is only partially correct. Which part(s) are correct?
   b. Which parts are incorrect?
   c. Give the correct five-number summary.
   d. What is the range of the class sizes? What is the interquartile range (IQR)?
   e. About ___% of math classes have 29 or more students in them.
   f. One-half of the math classes have more than ___ students.

Find the five-number summary for each data set.

5. 7, 14, 15, 19, 20, 25, 27

6. 40, 43, 45, 47, 54, 60

7. 52, 63, 66, 74, 80, 85, 91, 99

8. 4, 7, 7, 9, 10, 15, 26, 28, 34

Given the five-number summaries, find the interquartile range (IQR) for each data set.

9. 21 ~ 23 ~ 24 ~ 29 ~ 35

10. 16 ~ 28 ~ 34 ~ 37 ~ 40

11. 67 ~ 74 ~ 81 ~ 88 ~ 95

12. Wendy's pig had a litter of piglets. The five-number summary of the piglets' weights, in pounds, is given below.
    2.1 ~ 2.4 ~ 2.6 ~ 2.9 ~ 3.0
   a. Fifty percent of the piglets weighed less than or equal to ____ pounds.
   b. ____% of the piglets weighed 2.4 pounds or more.
   c. The middle 50% of the piglets weighed between ____ and ____ pounds.

Create a box-and-whisker plot for each five-number summary.

13. 5 ~ 8 ~ 10 ~ 18 ~ 29

14. 62 ~ 78 ~ 84 ~ 92 ~ 100

15. 49 ~ 59 ~ 62 ~ 64 ~ 72

Find the five-number summary for each set of data. Then create a box-and-whisker plot.

16. 48, 57, 63, 64, 66, 68, 68, 71

17. 11, 17, 7, 15, 14, 20, 14, 3, 19

18. The box-and-whisker plot shows the cost of tickets for the opera rounded to the nearest dollar. Use the information to answer the questions.
   a. The cheapest ticket to the opera costs ___ dollars.
   b. Half of the tickets cost less than ___ dollars.
   c. Twenty-five percent of the tickets cost more than ___ dollars.
   d. What is the range of the ticket prices?
   e. What is the interquartile range (IQR) of the ticket prices?

19. Kristen makes stuffed animals and sells them at the student store. She sells her stuffed animals at different prices based on the amount of material needed and the time they take to make. Make a box-and-whisker plot for the prices of Kristen's stuffed animals.

Cost of Stuffed Animals
$3, $6, $10, $10, $12, $14, $15, $15, $15, $16, $18, $20, $28, $35

REVIEW

20. A real estate agent collected data about the values of the last twelve homes she sold.

| $200,000 | $140,000 | $270,000 | $310,000 | $219,000 | $235,000 |
| $199,000 | $265,000 | $245,000 | $219,900 | $250,000 | $279,000 |

   a. Using an interval width of $50,000, draw a histogram of the data.
   b. Using an interval width of $20,000, draw a histogram of the data.
   c. Which histogram do you prefer? Explain your reasoning.
   d. What percent of the homes had a value greater than $250,000?

21. The dot plot below shows the number of servings of fruit eaten each week by a class of students.

Number of Fruit Servings Eaten Per Week

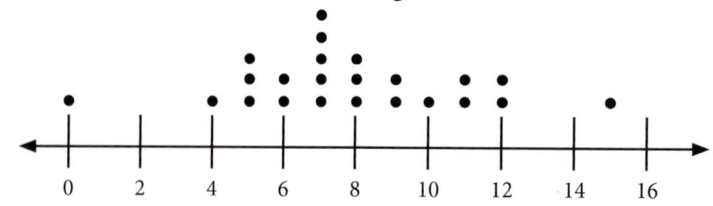

   a. Starting at 0 and using an interval width of 3, display this data in a histogram.
   b. In which interval does the 15 belong?
   c. Which graph (the dot plot or histogram) shows the clustering of the data better? Explain your reasoning.
   d. Which graph allows you to find the mode of the data set? Explain your reasoning.
   e. Which graph do you prefer for displaying this data set? Explain your reasoning.

146 Lesson 4.5 ~ Box-and-Whisker Plots

Tic-Tac-Toe ~ Prediction Time!

In this book, you have learned how to use rates to solve problems. You can apply these skills to solving problems involving histograms and box-and-whisker plots.

Use the histogram at right to answer #1 and #2.

1. If 120 sunflowers were collected, how many would you expect to be between 72 and 76 inches tall?

2. If 180 sunflowers were collected, how many would you expect to be less than 72 inches tall?

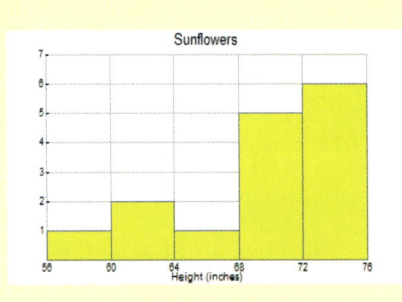

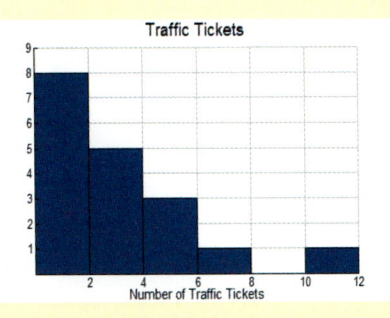

3. The histogram at left shows the number of traffic tickets received by people in a recent survey. Based on this survey, how many people in a group of 72 would likely have received 4 or 5 traffic tickets?

Use the box-and-whisker plot at right to answer #4 and #5.

4. Of the next 80 cars on this highway, how many would you expect to be driving more than 62 mph?

5. Of the next 120 cars on this highway, how many would you expect to be driving more than 48 mph?

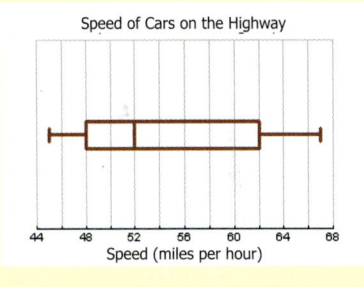

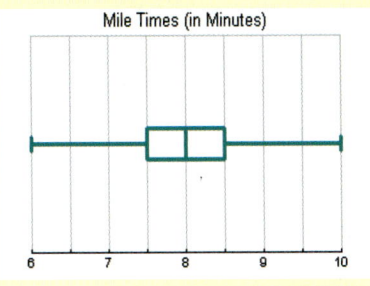

6. The box-and-whisker plot shows students' mile times in minutes. If a class of 36 students ran the mile, how many would you expect to run one mile in less than 8.5 minutes?

Lesson 4.5 ~ Box-and-Whisker Plots **147**

ANALYZING STATISTICS

LESSON 4.6

Analyze how characteristics of a data set affect the measures of center.

EXPLORE! **WHAT'S THE "MEAN"ING?**

Mr. Hinton was curious about the average number of letters in each of his students' names. The names of eight of his students are listed below.

Paul, Rob, Ana, Javon, Savannah, Ali, Juan, Alexandria

Step 1: Find the number of letters in each of the eight names.

Step 2: For each student, make a stack of cubes with the height matching the number of letters in his or her name. For example, Paul's stack should be 4 cubes tall.

Step 3: Look at the stacks. What is the mode?

Step 4: Put the stacks in order from shortest to tallest. Find the median.

Step 5: Without adding any more cubes, "level" the stacks by redistributing blocks so that all eight stacks have the same height. How many cubes are in each stack?

Step 6: If each of Mr. Hinton's eight students had the same number of letters in their name, how many letters would each person have? Which measure of center does this value represent?

Step 7: Which measure of center best represents the data? Explain your reasoning.

Step 8: Copy the dot plot below onto a sheet of paper. The numbers (instead of dots) on the top of the dot plot represent how far each value is away from the mean. These numbers are called absolute deviations from the mean. For example, Paul is represented by a "1" since his name has 4 letters – 1 away from the mean of 5. Rob and Ana are each represented by a "2" since they are 2 away from the mean. Which student does the "3" represent?

```
    2
    2    1    0              3
────┼────┼────┼────┼────┼────┼────┼────┼──▶
    3    4    5    6    7    8    9    10
         Number of Letters in Name
```

Each absolute deviation from the mean is written as a positive number.

Step 9: The dot plot is missing numbers representing Ali, Juan and Alexandria. Find where they should be located on the dot plot and write the absolute deviation number to represent each of them.

Step 10: Add the numbers to the left of the 0 on the dot plot. Then add the numbers to the right of the 0. What do you notice? What does this tell you about the mean as a "balancing" value for the data set?

148   Lesson 4.6 ~ Analyzing Statistics

When analyzing statistics, it is important to take the following into account: where the data came from, how much data was collected, how spread out the data is and any clusters that are present. In some cases one measure of center may be misleading and another helpful. An extreme value that varies greatly from the other values in a data set is called an **outlier**. Outliers can have a large impact on the mean, but little impact on the median or mode. Statistics from a larger data set are usually more reliable than statistics from a smaller data set.

EXAMPLE 1

Find the mean, median and mode of each data set. Determine which measure of center best represents each data set.
a. 4, 2, 2, 8, 6, 2, 8, 5, 8
b. 18, 19, 12, 17, 1, 19, 19
c. 10, 10, 7, 10, 10, 8, 10, 10, 10, 10

SOLUTIONS

a. Mean = $\frac{45}{9}$ = 5   Median = 5   Modes = 2 and 8

Since there are two modes, that would not be the best measure of center. The mean or median represents the data the best.

b. Mean = $\frac{105}{7}$ = 15   Median = 18   Mode = 19

The median best represents this data set. The mean is affected by the outlier (1) and the mode does not really represent the middle of the data set.

c. Mean = $\frac{95}{10}$ = 9.5   Median = 10   Mode = 10

All three measures of center represent this data set well. However, when there are many values that are the same, the mode is the best choice. The mode of 10 best represents this data set.

When comparing two sets of data, it is often helpful to analyze the data using a variety of statistics. Measures of center, ranges, five-number summaries and data displays can help in the analysis.

Lesson 4.6 ~ Analyzing Statistics

EXAMPLE 2

Jessie and Samantha compared how many movies they watched per month for a year. The two dot plots show each set of data:

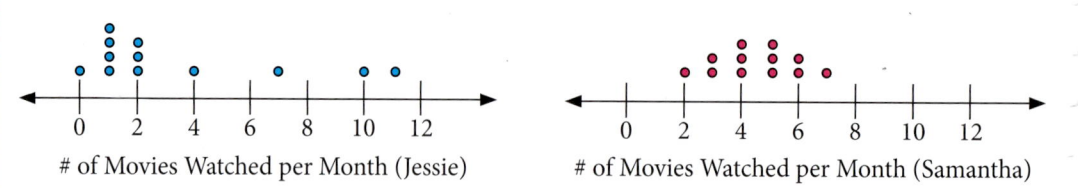

# of Movies Watched per Month (Jessie)

# of Movies Watched per Month (Samantha)

a. Compare the dot plots and describe the differences you see.
b. Find the measures of center for both Jessie and Samantha. How do they compare?
c. Find the range for each data set. How do they compare?
d. Which person is likely to watch more than 4 movies in a month? Which is more likely to watch more than 7 movies in a month? Explain your reasoning.

SOLUTIONS

a. Jessie's dot plot is more spread out than Samantha's. He has more low values, but also has some high numbers. Samantha's dot plot is symmetrical.
b. Jessie: Mean = 3.5; Median = 2; Mode = 1
Samantha: Mean = 4.5; Median = 4.5; Mode = 4, 5
Jessie's measures of center are lower than Samantha's.
c. Jessie's Range = 11 − 0 = 11; Samantha's Range = 7 − 2 = 5
Jessie's values are much more spread out.
d. Samantha is more likely to watch more than 4 movies in a month. Samantha did this six out of the twelve months compared to Jessie, who watched more than 4 movies only three times.
Jessie is more likely to watch more than 7 movies in a month. Samantha never watched more than 7 movies in a month.

EXERCISES

1. The number of hours worked per week by employees at a local restaurant are shown.

38, 45, 40, 42, 37, 8, 37, 41

  a. Find the mean, median and mode.
  b. Does there appear to be any outliers? Explain your reasoning.
  c. Which measure of center best represents this data?

2. Shruti tries to convince her parents that she deserves a larger allowance. She tells them that her classmates get an allowance of $20 per week on average.
  a. Shruti got her average by surveying eight of her friends. Describe the bias that may exist in this situation.
  b. The data that Shruti collected is shown. What is misleading about the average that Shruti used?

Classmates' Weekly Allowance

$12, $12, $30, $16, $15, $50, $15, $10

  c. Which measure of center would have been more appropriate to describe the typical allowance of Shruti's classmates? Explain your reasoning.

3. The weekly wages earned by two co-workers at Georgia-Mae's Café during the past seven weeks are shown below.

Tamiqua: $119, $121, $122, $125, $125, $127, $129
Jim: $112, $118, $119, $122, $129, $132, $136

  a. How do Tamiqua's mean weekly wage and Jim's mean weekly wage compare?
  b. Compare the median weekly wages of each employee.
  c. Which employee is most likely to earn at least $120 next week? Explain your reasoning.
  d. Which employee is most likely to earn at least $130 next week? Explain your reasoning.

4. Simone has scored 82, 74, 94, 74, 78 and 96 on her English tests this term.
  a. She wants to convince her parents that she is doing well in the class. Which measure of center should Simone use as her evidence? Explain your reasoning.
  b. Simone's teacher shows her that she has room to improve. Which measure of center did her teacher use to show that Simone has room for improvement? Explain your reasoning.

5. Travis and Alden just finished a three-game bowling match. A disagreement about who won occurred. Use the scores and statistics to settle the dispute.

	Game 1	Game 2	Game 3
Travis	175	129	152
Alden	148	169	151

  a. Make a case for why Travis should be declared the winner.
  b. Make a case for why Alden should be declared the winner.
  c. In your opinion, who should be declared the winner of the bowling match? Use the statistical evidence to determine your choice. Explain your reasoning.

6. Mr. Tobin and Mrs. Vicente compared their students' scores on their latest quiz. The results are shown below.

Mr. Tobin: 8, 6, 9, 1, 3, 10, 5, 1, 7, 9, 2, 10
Mrs. Vicente: 3, 7, 9, 2, 7, 6, 1, 4, 10, 7, 8, 4

  a. Make a histogram for each teacher's scores. Start at 0 and use an interval width of 2.
  b. Describe the differences in the two histograms.
  c. Find the measures of center for each teacher. How do they compare?
  d. Find the five-number summary for each teacher.
  e. How does the IQR for Mr. Tobin's class compare to the IQR for Mrs. Vicente's class? What does this tell you about the spread of their data?

7. Which representations of data below can be used to calculate measures of center exactly?

Histogram   Box-and-Whisker Plot
Dot Plot   Frequency Table
Five-Number Summary   List of Values

8. For each part, create a display that shows the given information. Explain how you know your display is correct.
   a. The median and mean have the same value.
   b. The mode of the data is 10.
   c. The median of the data is 6.
   d. The outlier of the data is 20.

9. Gloria collected data about the number of pets that her neighbors had. Her results were: 1, 1, 3, 0, 2, 0, 1, 7, 3, 0, 4, 2
   a. Find the mean of her data.
   b. Make a dot plot similar to the one in **Step 8** of the **Explore!**. Instead of dots, use numbers to show the absolute deviations from the mean. For example, use 0s for dots on the mean value, use 1s for values that are 1 unit away from the mean, etc.
   c. Sum the numbers on your dot plot to the left of the mean. Then sum the numbers on your dot plot to the right of the mean. What should be true about these values?

REVIEW

Find the five-number summary for each set of data. Then create a box-and-whisker plot.

10. 7, 13, 15, 19, 19, 22, 26, 30, 32

11. 26, 39, 42, 18, 49, 30, 35, 50, 29, 46

12. A teacher surveyed her students: "In how many other states have you lived?" Use the box-and-whisker plot to answer the questions below:
   a. What is the median number of other states in which students have lived?
   b. Seventy five percent of her students have lived in more than ___ other states.
   c. What percent of her students have lived in somewhere between 1 and 6 other states?

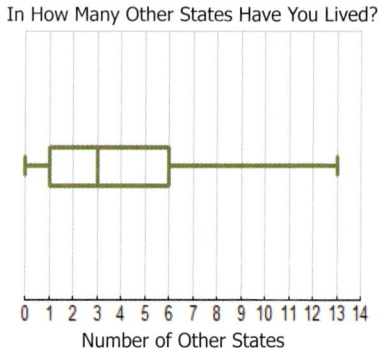

13. Describe the possible bias in the following survey situation:
   To collect data about 6th grade students' favorite musical groups, a student asks 20 of his 6th grade friends.

Tic-Tac-Toe ~ How Far Is Too Far?

Outliers were defined in **Lesson 4.6** as values that vary greatly from most of the other values in a data set. Unfortunately, this definition can cause disagreement about what qualifies as an outlier. Statisticians have numerous ways for determining whether a number is an outlier or not. One common method for determining outliers is called the IQR Method.

> ### The IQR Method
>
> **Step 1:** Find the IQR of the data.
> **Step 2:** Multiply 1.5 · IQR.
> **Step 3:** Find the sum of Q3 and your answer from Step 2: Q3 + (1.5 · IQR).
> This is the "upper boundary". Any numbers above this upper boundary are outliers.
> **Step 4:** Find the difference of Q1 and your answer from Step 2: Q1 − (1.5 · IQR).
> This is the "lower boundary". Any numbers below this lower boundary are outliers.

Use the given Q1 and Q3 values to determine the upper and lower boundaries for outliers.

1. Q1 = 22 and Q3 = 28

2. Q1 = 5 and Q3 = 10

Use the IQR Method to determine if the following sets of data have any outliers. If so, state the outliers. If there are none, state "no outliers".

3. 3, 7, 7, 10, 10, 11, 16

4. 16, 21, 21, 21, 22, 22, 24, 28

5. Wesley mows lawns as his summer job. The list shows the number of lawns he mowed each week for nine weeks during the summer.

Number of Lawns mowed: 3, 7, 16, 6, 6, 10, 8, 9, 8

The week of July 4th was his busiest when he mowed 16 lawns. Was that number an outlier? Use the IQR Method to justify your answer.

6. Dick Fosbury was said to have revolutionized the art of the high jump when he introduced the "Fosbury Flop" and won the gold medal in the 1968 Olympics. Was his 1968 jump an outlier? Show all work necessary to justify your answer.

Year	1920	1924	1928	1932	1936	1948	1952	1956	1960	1964	1968
Height (inches)	76.25	78	76.38	77.63	79.94	78	80.32	83.25	85	85.75	88.25

MEAN ABSOLUTE DEVIATION

LESSON 4.7

Find and use the mean absolute deviation to describe the spread of data.

In this block you've learned how to use measures of center, such as the mean, to summarize the values in a data set. You've also learned how to use range, IQR, data displays and five-number summaries to describe the spread of a data set. Another way to measure the spread of a data set is to see how far each number is away from the mean of the data set. The average of these distances is called the **mean absolute deviation**.

EXPLORE!   MERCURY'S RISING

Jeron and Gary are pen pals living across the United States from each other; Jeron lives in San Francisco, CA and Gary lives in Boston, MA. Gary wondered how the temperatures in the two cities compared to each other, so he looked up the average monthly high temperatures for each city. These temperatures are listed below:

San Francisco: 57, 57, 60, 62, 63, 63, 64, 67, 67, 68, 69, 70
Boston: 36, 39, 41, 45, 52, 56, 61, 66, 72, 76, 80, 82   Source: *weather.com*

Step 1: How do the highest and lowest average monthly temperatures in each city compare to each other?

Step 2: Find the mean temperatures for each city. Round to the nearest whole degree. Your rounded mean for San Francisco should match the value in the table below.

Step 3: Copy the Boston table below and write the mean value from **Step 2** in the blank.

San Francisco
Mean = 64

Average Monthly Temperature	Deviation from Mean	Absolute Deviation
57	57 − 64 = −7	7
57	57 − 64 = −7	7
60	60 − 64 = −4	4
62	62 − 64 = −2	2
63	63 − 64 = −1	1
63	63 − 64 = −1	1
64	64 − 64 = 0	0
67	67 − 64 = 3	3
67	67 − 64 = 3	3
68	68 − 64 = 4	4
69	69 − 64 = 5	5
70	70 − 64 = 6	6

Total = 43
Mean Absolute Deviation = 3.58

Boston
Mean = 59

Average Monthly Temperature	Deviation from Mean	Absolute Deviation
36	36−59=−23	23
39	39−59=−20	20
41	41−59=−18	18
45	45−59=−14	14
52	52−59=−7	7
56	56−59=−3	3
61	61−59=2	2
66	66−59=7	7
72	72−59=13	13
76	76−59=17	17
80	80−59=21	21
82	82−59=23	23

Total = 168
Mean Absolute Deviation = 14

43 ÷ 12 = 3.58

EXPLORE!   CONTINUED

Step 4: Find the "Deviations from the Mean" for Boston. Subtract the mean for Boston from each value in the Boston table (Mean – Data Value). Some of these values should be negative. Record these numbers in the "Deviation from Mean" column for Boston. See the San Francisco table as an example.

Step 5: Find the "Absolute Deviation" for each value in the Boston table. The absolute deviation is the absolute value of the deviation in the second column. Absolute deviations are always positive.

Step 6: Find the sum of all the "Absolute Deviations" in the Boston table. Finally, divide this sum by the number of data values (12) to find the mean absolute deviation. Round to two decimal places.

Step 7: How does the mean absolute deviation of San Francisco's temperatures compare to that of Boston? What does this tell you about the variations in the temperatures of the two cities?

HOW TO FIND THE MEAN ABSOLUTE DEVIATION OF A DATA SET

1. Find the mean of the data set.
2. Find the deviation from the mean for each data value.
          Data Value – Mean
3. Find the absolute deviation for each data value. These are always positive.
4. Find the mean of the absolute deviations.

The smaller the mean absolute deviation, the less spread there is in the data set. It means the numbers are all close to the mean. When the mean absolute deviation is larger, the numbers are more spread out from the mean and there is greater variability in the numbers in the data set.

EXAMPLE 1

Bonnie surveyed six classmates about the number of hours of TV they watched each week.   6, 7.5, 4, 10, 13.5, 7

a. What is the mean absolute deviation of the data set?
b. What does this value mean in terms of the data set?

SOLUTIONS

a. Find the mean of the data set. $\dfrac{4 + 6 + 7 + 7.5 + 10 + 13.5}{6} = \dfrac{48}{6} = 8$

Organize the work in a table. Write the data values from Bonnie's survey in order to keep the information more organized.

For each value, find the deviation from the mean and the absolute deviation.

Average Hours TV Watching	Deviation from Mean	Absolute Deviation
4	4 – 8 = –4	4
6	6 – 8 = –2	2
7	7 – 8 = –1	1
7.5	7.5 – 8 = –0.5	0.5
10	10 – 8 = 2	2
13.5	13.5 – 8 = 5.5	5.5

Lesson 4.7 ~ Mean Absolute Deviation

EXAMPLE 1 SOLUTIONS (CONTINUED)

Find the mean of the absolute deviations. $\frac{4+2+1+0.5+2+5.5}{6} = \frac{15}{6} = 2.5$

The mean absolute deviation is 2.5.

b. On average, the number of hours students watch TV per week is within about 2.5 hours of the mean (8 hours).

EXERCISES

1. How is the mean absolute deviation different than the mean of a data set?

2. How is the mean absolute deviation different than the interquartile range (IQR) of a data set?

3. Use the mean stated in the table below to find the deviations from the mean for each value.

					Mean = 12
Data Values	5	10	13	15	17
Deviation from the Mean					

4. In **Exercise 3** above, what are the **absolute** deviations from the mean?

5. Follow the steps below to find the mean absolute deviation of the following data set:

Ages of Used Cars in a Car Lot: 2, 5, 8, 9, 10, 10, 12

a. Find the mean of the data.
b. Find the deviations from the mean for each of the values.
c. Find the absolute deviations from the mean.
d. Find the mean absolute deviation.

6. Rodrigo and Glenn kept track of the money (in dollars) they earned mowing lawns this year.

	April	May	June	July	August	September
Rodrigo	40	55	40	20	20	25
Glenn	35	35	30	30	25	35

a. Find the mean of Rodrigo's monthly earnings.
b. Find the mean absolute deviation of Rodrigo's monthly earnings. Create a table like **Example 1** to organize your work.
c. Find the mean of Glenn's monthly earnings.
d. Find the mean absolute deviation of Glenn's monthly earnings. Use a table to organize your work.
e. Compare the results. Write two sentences to explain the meaning of your answers from **parts a-d**.

7. Coach Ryland wants his football team to have consistent passing yards. The team's goal is to average more than 250 passing yards per game and to not have a mean absolute deviation of more than 20 yards. The table shows the team's passing yards the past nine weeks.

	Week 1	Week 2	Week 3	Week 4	Week 5	Week 6	Week 7	Week 8	Week 9
Passing Yards	280	250	280	260	210	240	280	300	240

    a. Find the mean of the data. Did the team meet their goal of more than 250 passing yards?
    b. Find the mean absolute deviation. Did the team meet their goal of a mean absolute deviation of less than 20 yards?

8. The heights (in inches) of Mr. Ebler's students are shown in the dot plot below. Use the plot to answer the following questions.

Heights of Mr. Ebler's Students (inches)

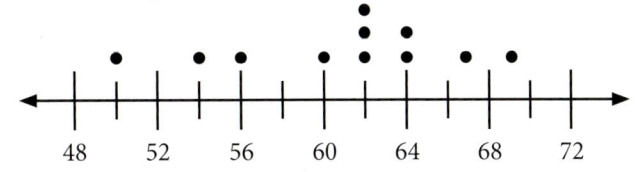

[dot plot with values between 48 and 72]

    a. Find the mean of the data. Round to the nearest whole number.
    b. Using the rounded mean from **part a,** find the mean absolute deviation.
    c. Find the five-number summary of the data.
    d. What is the IQR of the data set?
    e. How does the mean absolute deviation compare to the IQR?

9. Create a data set with 7 values in which the mean absolute deviation is larger than the IQR.

10. Consider the following data set: 3, 6, 7, 11, 13
    a. Find the mean absolute deviation of the data set.
    b. Replace the 13 in the data set with 33. Find the new mean absolute deviation.
    c. Do outliers have an impact on the mean absolute deviation of a data set? Explain your reasoning.

REVIEW

11. Sid kept track of his hours worked for the last fourteen weeks and put them in the table.

Hours Worked

22	25.5	32	20	19.5	21	24
19	26	8	23	20.5	24	25

    a. Find the minimum and maximum values in the data set. What would be a reasonable interval width to use for this data set?
    b. Use your interval width in **part a** to create a frequency table for the data. Be sure that each data value is only included in one interval.
    c. Use your frequency table to create a histogram. Be sure to label both axes.

12. Brenton went fishing with his dad each of the last 6 weekends. He caught the following number of fish:

$$1, 0, 2, 9, 0, 1$$

　a. Find the mean, median and mode of the data.

　b. Do there appear to be any outliers? Explain your reasoning.

　c. Which measure of center best represents how many fish Brenton can expect to catch when he goes fishing? Explain your reasoning.

Which measure of center best represents the following data sets? Explain your reasoning.

13. 6, 8, 8, 8, 8, 8, 9, 10

14. 13, 14, 14, 15, 18, 19, 20

Tic-Tac-Toe ~ Which Graph Should I Use?

In this block, you have learned how to make a variety of graphs. Some graphs work better for certain situations or data sets.

For each of the situations below, choose which type of graph (bar graph, dot plot, histogram or box-and-whisker plot) would be best to use to display the data set. Use each type of graph at least once and justify your choice for each.

1. the heights of students in your class

2. the favorite food of students in your class

3. the shoe sizes of twelve players on the basketball team

4. the number of televisions that students have in their house

5. the maximum speeds of the top 20 roller coasters in the United States

6. the prices of 40 new cars at a local car dealership

Make an appropriate graph for each set of data. Use each type of graph below only once.

　　　Histogram　　　Box-and-Whisker Plot　　　Dot Plot　　　Bar Graph

7. Speed of Cars on the Interstate (*mph*)
58, 59, 61, 64, 65, 65, 67, 68, 69, 72, 74

8. Number of Amusement Parks Students Have Been To
0, 0, 1, 2, 3, 3, 3, 5, 6, 6, 8, 9, 12, 14, 15

9. Costs of Concert Tickets in Dollars
25, 28, 29, 30, 32, 35, 39, 40, 40, 42, 42, 45, 45, 45, 48, 48, 52, 54, 56, 67, 72, 72, 75, 89, 95

10.

How Time is Spent on the Internet				
Finances	Research	Entertainment	Shopping	Communication
10%	15%	25%	15%	35%

158　Lesson 4.7 ~ Mean Absolute Deviation

TIC-TAC-TOE ~ THE BIG AND SMALL OF IT

Answer the following questions to investigate how outliers affect large and small data sets.

Heidi is a strong student in her art class. Here are her grades on the first three tests of the term:

89%, 92%, 88%

1. What is Heidi's mean test score after three tests?

2. Heidi is really concerned about the upcoming fourth test. She wonders if she earns 20%, how much it would affect her grade. If Heidi does earn 20%, what would be her new mean?

3. How much would her mean drop between #1 and #2?

Heidi does well on her fourth test. In fact, she does well on her next several tests. Here are her tests to this point in the term:

89%, 92%, 88%, 89%, 92%, 88%, 89%, 92%, 88%

4. What is Heidi's current mean on the nine tests this term?

5. The tenth test of the term is coming up and this time Heidi is really concerned and is convinced she's not going to do well. If Heidi earns 20% on the tenth test, what would be her new mean?

6. How much would her mean drop between #4 and #5?

7. According to your answers on #3 and #6, does an outlier have a larger effect on a small data set (data set with a small amount of numbers) or a large data set (data set with a large amount of numbers)? Explain your reasoning.

For each problem below, find the mean of both original data sets. Then find the new mean for each data set after the indicated change is made.

8. Set A: 19, 23, 21, 14, 23
   Set B: 21, 16, 23
   Change: Include the number 40 in each data set

9. Set A: 19, 26, 30, 33
   Set B: 24, 25, 25, 26, 26, 27, 36
   Change: Include the number 4 in each data set

10. In each problem above, which data set's mean changed more when the new number was included? Does this confirm what you observed in #7 above?

Lesson 4.7 ~ Mean Absolute Deviation 159

REVIEW

BLOCK 4

Vocabulary

1st quartile (Q1)	five-number summary	median
3rd quartile (Q3)	frequency table	mode
box-and-whisker plot	histogram	numerical data
bias	interquartile range (IQR)	outlier
categorical data	mean	range
dot plot	mean absolute deviation	statistics
	measures of center	

Identify and write statistical questions and the type of data that will result.
Find measures of center and range.
Display data using dot plots.
Make, read and interpret histograms.
Make, read and interpret box-and-whisker plots.
Analyze how characteristics of a data set affect the measures of center.
Find and use the mean absolute deviation to describe the spread of data.

Lesson 4.1 ~ Introduction to Statistics

Classify each data set as either categorical or numerical.

1. the time of day that students go to bed

2. the month that students were born in

3. the number of bowls of cereal you eat per month

4. students' favorite type of cereal

For each problem below, determine which is a better statistical question. Explain your reasoning.

5. A. *What is the most popular brand of shoe in the United States?* **OR**
   B. *How many pairs of shoes do you own?*

6. A. *How many glasses of orange juice do you drink per week?* **OR**
   B. *Do you like to drink orange juice?*

For each surveying situation, describe how there could be some bias. Explain how each could be changed to eliminate the bias.

7. Surveying customers outside of a pizza parlor to determine how many times per month people eat pizza.

8. To determine whether people liked the food at a restaurant, handing out mail-in questionnaires to people leaving the restaurant.

9. The graph at left shows the change in the cost of a digital camera over one year. What is misleading about the graph? Draw a new graph that is not misleading.

Lesson 4.2 ~ Measures of Center

Find the three measures of center. Round answers to the nearest tenth, if necessary.

10. 12, 14, 14, 24, 26

11. 10, 14, 14, 18, 23, 23

12. 9, 25, 4, 13, 24, 30, 17

13. 38, 41, 29, 46, 49, 40, 33, 50, 43, 38

14. April has earned $30, $50 and $45 the last three months babysitting. How much does she need to earn the next month in order to average $45 per month? Show all work necessary to justify your answer.

Find the range of the following sets of data.

15. 39, 45, 47, 52, 55

16. 38, 32, 32, 28, 37, 31

Use the range and the mean to determine the missing numbers in each ordered data set. Show all work necessary to justify your answers.

17. __ , 11, __ , 18, 20
Mean = 14
Range = 14

18. 2, 4, 8, ___ , 17, 17, 21, ___
Mean = 13
Range = 21

Lesson 4.3 ~ Dot Plots

Use the dot plot to answer the following questions.

19. How many students were surveyed?

20. What is the mode of the number of fish that students have?

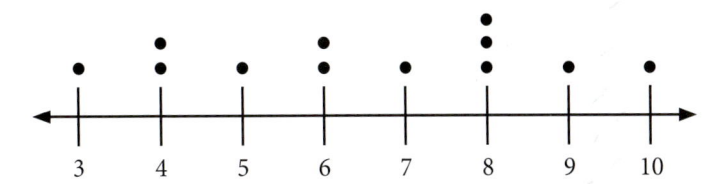

21. What is the median number of fish that students have?

22. What percent of students have at least 5 fish in their aquarium? Use mathematics to justify your answer.

23. At the latest track and field meet, fifteen students threw the shot put. The top distances (in feet) of each of their throws are shown.
   a. Sketch a dot plot of the data.
   b. Describe the spread of the data in the plot.

30	32	26	34	40
31	32	32	35	25
34	33	32	33	30

Lesson 4.4 ~ Histograms

24. How does a histogram differ from a bar graph?

25. An ideal histogram has between ___ and ___ intervals.

26. Use the histogram below to answer the following questions.

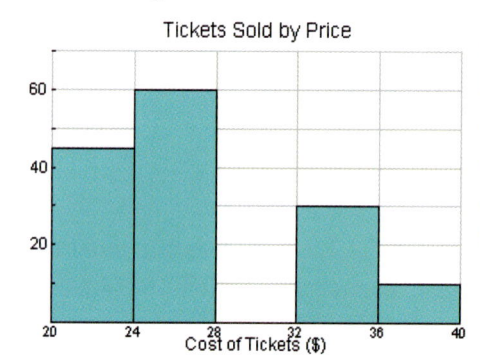

   a. What is the interval width?
   b. How many total tickets were sold?
   c. Which price interval sold the most tickets?
   d. If a ticket were sold for $24, in which interval would it fall?
   e. How many tickets were sold for less than $28?
   f. The 28-32 interval is empty. What does that mean?

27. Sara was comparing the cost of a large coffee at several coffee shops. She went to fourteen different coffee shops and recorded their prices.

Price of a 20-Ounce Coffee						
$1.10	$1.49	$2.15	$1.00	$1.15	$1.25	$1.50
$1.79	$1.59	$1.75	$1.15	$1.30	$1.50	$1.49

   a. Find the minimum and maximum values in the data set. What would be a reasonable interval width to use for this data set? Explain your reasoning.
   b. Use your interval width in **part a** to create a frequency table for the data. Be sure that each data value is only included in one interval.
   c. Use your frequency table to create a histogram. Be sure to label both axes.

Lesson 4.5 ~ Box-and-Whisker Plots

28. Use words or a picture to describe how the five-number summary is used to create a box-and-whisker plot.

29. The five-number summary breaks a data set down into quartiles where each section represents ___% of the data.

Find the five-number summary for each data set.

30. 22, 29, 30, 32, 36, 36, 40

31. 19, 6, 4, 20, 9, 25, 11, 15, 16

For each five-number summary below:
a. Create a box-and-whisker plot.
b. Find the interquartile range (IQR).

32. 20 ~ 32 ~ 41 ~ 46 ~ 50

33. 50 ~ 57 ~ 62 ~ 66 ~ 74

34. The list shows the "talk-time" battery life of ten cell phones offered by MotoPhone.

Talk Time (in hours)
MotoPhone: 10, 7, 16, 18, 13, 4, 8, 20, 15, 15

a. Make a box-and-whisker plot to display the battery lives of MotoPhone cell phones.
b. MotoPhone's median battery life is ____ hours.
c. Twenty-five percent of MotoPhone's phones have a battery life of at least ____ hours.
d. Seventy-five percent of MotoPhone's phones have a battery life of at least ____ hours.

Lesson 4.6 ~ Analyzing Statistics

35. The heights of several players on the basketball team are shown below.
Height in inches: 65, 67, 76, 62, 65, 67, 66
a. Find the mean, median and mode of the data.
b. Do there appear to be any outliers? Explain.
c. Which measure of center best represents this data? Explain your reasoning.

Which measure of center would best describe each data set? Explain your reasoning.

36. 3, 7, 8, 8, 8, 8, 8, 8, 9, 10

37. 20, 17, 12, 30, 14, 16, 18

38. Students in Mrs. Halper's and Mr. Giver's classes are doing a canned food drive. The lists below show the number of cans that each student in that class has contributed.
Mrs. Halper: 71, 60, 48, 37, 85, 111, 45, 70, 48, 55, 38, 44
Mr. Giver: 30, 22, 36, 35, 240, 12, 10, 270, 62, 28, 5, 17
a. What is the mean number of cans that was contributed by students in each class?
b. What is misleading about the means in each class?
c. Which class do you think has done a better job of contributing to the canned food drive? Explain your reasoning.

39. The dot plot below shows the number of countries visited outside the United States by Mr. Parker's students. Instead of dots, the plot shows the absolute deviations from the mean.

Number of Countries Visited Outside U.S.

```
              1
              1   0
          2   1   0
          2   1   0   1   2           5
      ←———+———+———+———+———+———+———+———→
          0   1   2   3   4   5   6   7
```

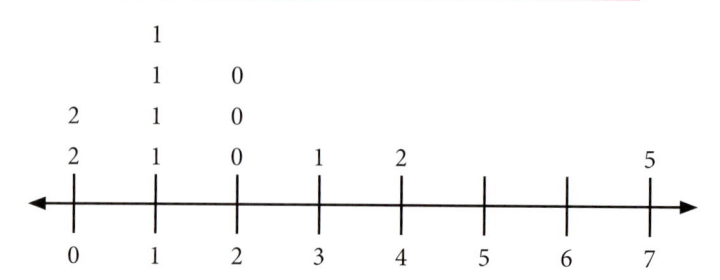

  a. How many students are included in the dot plot?
  b. What is the mode of the data?
  c. What is the mean of the data?
  d. Sum the numbers on the dot plot to the left of the mean. Then sum the numbers on the dot plot to the right of the mean. What do you notice about these values?

Lesson 4.7 ~ Mean Absolute Deviation

40. Use the mean stated in the table below to find the deviations from the mean for each value.

					Mean = 9
Data Values	2	8	10	11	14
Deviation from the Mean					

41. In **Exercise 40** above, what are the absolute deviations from the mean? Find the mean absolute deviation for the data set in **Exercise 40**. Show all work necessary to justify your answer.

42. The dot plot below shows the absolute deviations from the mean for the number of countries students have visited outside the United States. What is the mean absolute deviation of the data set? Use mathematics to justify your answer.

Number of Countries Visited Outside U.S.

```
              1
              1   0
          2   1   0
          2   1   0   1   2           5
      ←———+———+———+———+———+———+———+———→
          0   1   2   3   4   5   6   7
```

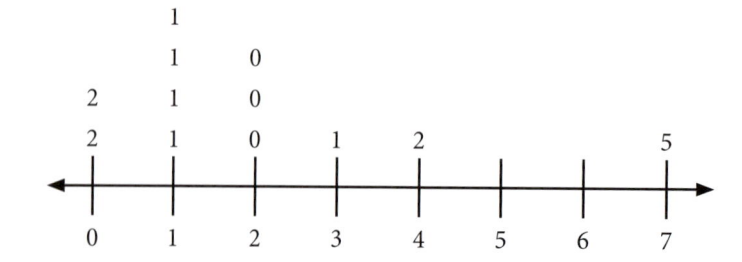

43. Chris likes to go on weekend bike rides across the state. Listed below are the lengths of his bike rides the past six weeks.

Length of bike ride (miles): 30, 34, 20, 44, 34, 30

  a. Find the mean distance of Chris' weekly bike rides.
  b. Find the absolute deviations from the mean for each of his bike rides.
  c. What is the mean absolute deviation of Chris' weekly bike rides?
  d. Write a sentence describing the meaning of your answer to **part c**.

Tic-Tac-Toe ~ Double the Whiskers

Creating parallel box-and-whisker plots can be helpful in comparing two or more data sets. These can be created by placing one box-and-whisker plot above another on the same number line.

To create a parallel box-and-whisker plot, draw and label a number line that will include both the minimum and maximum values of both data sets. Then use the five-number summaries to create two box-and-whisker plots. Place the plots one above the other and label each.

Complete the following problems.

1. Two different fertilizers were given to several plants of the same type. The parallel box-and-whisker plot shows the distribution of the heights (in inches) of the plants after 6 weeks.
  a. What is the five-number summary for each fertilizer?
  b. What are the advantages of each fertilizer?
  c. Which fertilizer is more likely to produce a plant that is at least 14 inches in height? Explain your reasoning.
  d. Which fertilizer is more likely to produce a plant that is at least 22 inches in height? Explain your reasoning.

2. Nicole and Fatima were comparing the costs of jeans at their two favorite stores, Denim Factory and Dream Jeanie's. They randomly grabbed jeans of various styles at each store and recorded their prices as follows:

Prices ($) at Denim Factory	50	30	35	60	45	50	30	38	40	40
Prices ($) at Dream Jeanie's	55	50	50	28	50	55	18	58	25	45

  a. Make a parallel box-and-whisker plot to compare the prices at the two stores.
  b. Nicole says that Denim Factory has better prices. Fatima prefers Dream Jeanie's prices. Which store do you think has better prices? Explain your reasoning.

CAREER FOCUS

Meghan
ASL Interpreter

I am an American Sign Language Interpreter. I interpret for hearing and deaf people who need to communicate with each other using English and American Sign Language. The things I hear in English, I sign, and the things that are signed to me, I speak in English.

ASL Interpreters can work in all kinds of settings, such as schools, hospitals, courts and businesses. An American Sign Language interpreter can also do video interpreting and interpret between a person speaking on the phone and a person signing through video. Any place a deaf person may go, an interpreter could work there.

I use math every day when I interpret for schools, whether it is elementary, middle school, high school or college. I need to have a solid background in math so that I can explain concepts clearly through sign language. I use many different kinds of math: geometry, algebra, trigonometry, calculus and more in my profession. Interpreters also use math when interpreting in business and medical settings.

In order to become an American Sign Language Interpreter, I earned a Bachelor's degree in American Sign Language/English Interpreting. Some people choose to earn an Associate's degree before working in this field. In addition to any degree earned, you must also be fluent in American Sign Language and have an interpreter certification.

The average ASL interpreter earns $45,700. An educational interpreter's salary can start at about $20,000 per year, while interpreters working in government or business can earn up to $75,000 per year.

I love having a job that is so unique. I like helping others and helping to ease communication issues between people. I enjoy having an opportunity to educate people about challenges to those who are deaf and hard of hearing and to also teach people about deaf culture.

ACKNOWLEDGEMENTS

All Photos and Clipart ©2008 Jupiterimages Corporation and Clipart.com with the exception of the cover photos and the following photos:

Ratios, Rates & Statistics Page 6
©iStockphoto.com/Orlando Rosu

Ratios, Rates & Statistics Page 7
©iStockphoto.com/Simon Ingate

Ratios, Rates & Statistics Page 12
©iStockphoto.com/sjlocke

Ratios, Rates & Statistics Page 14
©iStockphoto.com/andipantz

Ratios, Rates & Statistics Page 16
©iStockphoto.com/Mark Herreid

Ratios, Rates & Statistics Page 22
©iStockphoto.com/YinYang

Ratios, Rates & Statistics Page 25
©iStockphoto.com/Jaap Hart

Ratios, Rates & Statistics Page 29
©iStockphoto.com/Okea

Ratios, Rates & Statistics Page 37
©iStockphoto.com/Tomasz Zagórski

Ratios, Rates & Statistics Page 38
©iStockphoto.com/Ron Bailey

Ratios, Rates & Statistics Page 39
©iStockphoto.com/tawnie cleric

Ratios, Rates & Statistics Pge 41
©iStockphoto.com/George Clerk

Ratios, Rates & Statistics Page 41
©iStockphoto.com/Alexander Belyaev

Ratios, Rates & Statistics Page 45
©iStockphoto.com/DundStock

Ratios, Rates & Statistics Page 46
©iStockphoto.com/steve greer

Ratios, Rates & Statistics Page 51
©iStockphoto.com/Alan Eisen

Ratios, Rates & Statistics Page 51
©iStockphoto.com/james steidl

Ratios, Rates & Statistics Page 54
©iStockphoto.com/sumnersgraphicsinc

Ratios, Rates & Statistics Page 55
©iStockphoto.com/Henrik Jonsson

Ratios, Rates & Statistics Pge 57
©iStockphoto.com/Michael Flippo

Ratios, Rates & Statistics Page 58
©iStockphoto.com/Lisa Denise Hillström

Ratios, Rates & Statistics Page 63
©iStockphoto.com/Christine Gonsalves

Ratios, Rates & Statistics Pge 68
©iStockphoto.com/olaf herschbach

Ratios, Rates & Statistics Page 69
©iStockphoto.com/Pavel Lebedinsky

Ratios, Rates & Statistics Page 69
©iStockphoto.com/johnnyscriv

Ratios, Rates & Statistics Pge 73
©iStockphoto.com/Amanda Rohde

Ratios, Rates & Statistics Page 76
©iStockphoto.com/ShutterWorx

Ratios, Rates & Statistics Page 77
©iStockphoto.com/poligonchik

Ratios, Rates & Statistics Page 80
©iStockphoto.com/Larry Masseth

Ratios, Rates & Statistics Page 83
©iStockphoto.com/Vitali Dyatchenko

Ratios, Rates & Statistics Page 85
©iStockphoto.com/Tibor Nagy

Ratios, Rates & Statistics Page 86
©iStockphoto.com/Toru Uchida

Ratios, Rates & Statistics Page 87
©iStockphoto.com/Maksym Bondarchuk

Ratios, Rates & Statistics Page 87
©iStockphoto.com/Csondy

Ratios, Rates & Statistics Pge 88
©iStockphoto.com/Douglas Litchfield

Ratios, Rates & Statistics Page 90
©iStockphoto.com/xril

Ratios, Rates & Statistics Page 91
©iStockphoto.com/Nuno Silva

Ratios, Rates & Statistics Pge 93
©iStockphoto.com/svetlana foote

Ratios, Rates & Statistics Page 95
©iStockphoto.com/gary milner

Ratios, Rates & Statistics Page 96
©iStockphoto.com/Holly Kuchera

Ratios, Rates & Statistics Page 97
©iStockphoto.com/David Lee

Ratios, Rates & Statistics Page 97
©iStockphoto.com/james steidl

Ratios, Rates & Statistics Page 101
©iStockphoto.com/manfredxy

Ratios, Rates & Statistics Page 103
©iStockphoto.com/Carol Oostman

Ratios, Rates & Statistics Page 106
©iStockphoto.com/Jodi Jacobson

Ratios, Rates & Statistics Page 108
©iStockphoto.com/OGphoto

Ratios, Rates & Statistics Page 111
©iStockphoto.com/Lisa F. Young

Ratios, Rates & Statistics Page 111
©iStockphoto.com/Pablo Caridad

Ratios, Rates & Statistics Page 112
©iStockphoto.com/Okea

Ratios, Rates & Statistics Page 112
©iStockphoto.com/zentilia

Ratios, Rates & Statistics Page 117
©iStockphoto.com/VINCENT GIORDANO

Ratios, Rates & Statistics Page 120
Scott Valway

Ratios, Rates & Statistics Page 121
©iStockphoto.com/Anne-Louise Quafoth

Ratios, Rates & Statistics Page 121
©iStockphoto.com/Alexey Dudoladov

Ratios, Rates & Statistics Page 125
©iStockphoto.com/DRB Images, LLC

Ratios, Rates & Statistics Page 125
©iStockphoto.com/Sergii Figurnyi

Ratios, Rates & Statistics Page 129
©iStockphoto.com/Sandra Gligorijevic

Ratios, Rates & Statistics Page 130
©iStockphoto.com/Ismail Akin Bostanci

Ratios, Rates & Statistics Page 130
©iStockphoto.com/hulyaguzel

Ratios, Rates & Statistics Page 131
©iStockphoto.com/Don Bayley

Ratios, Rates & Statistics Page 132
Scott Valway

Ratios, Rates & Statistics Page 137
Scott Valway

Ratios, Rates & Statistics Page 138
©iStockphoto.com/adam fraser

Ratios, Rates & Statistics Page 139
©iStockphoto.com/cynoclub

Ratios, Rates & Statistics Page 141
©iStockphoto.com/kycstudio

Ratios, Rates & Statistics Page 145
©iStockphoto.com/Ilka-Erika Szasz-Fabian

Ratios, Rates & Statistics Page 148
©iStockphoto.com/Christopher Futcher

Ratios, Rates & Statistics Page 149
©iStockphoto.com/Jacob Wackerhausen

Ratios, Rates & Statistics Page 149
©iStockphoto.com/Catherine Yeulet

Ratios, Rates & Statistics Page 151
©iStockphoto.com/Steve Debenport

Ratios, Rates & Statistics Page 151
©iStockphoto.com/Isselée

Ratios, Rates & Statistics Page 152
©iStockphoto.com/Jason Lugo

Ratios, Rates & Statistics Page 156
©iStockphoto.com/egeeksen

Layout and Design by Judy St. Lawrence

Cover Design by Schuyler St. Lawrence

Glossary Translation by Keyla Santiago and Heather Contreras

CORE FOCUS ON MATH
GLOSSARY ~ GLOSARIO

A

Absolute Value	The distance a number is from 0 on a number line.	**Valor Absoluto**	La distancia de un número desde el 0 en una recta numérica.
Acute Angle	An angle that measures more than 0° but less than 90°. 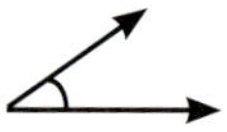	**Ángulo Agudo**	Un ángulo que mide mas 0° pero menos de 90°. 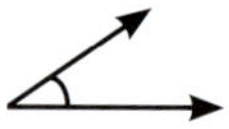
Adjacent Angles	Two angles that share a ray. 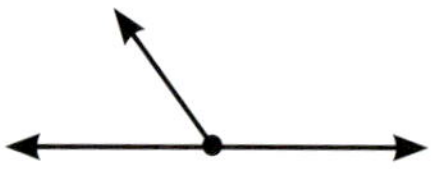	**Ángulos Adyacentes**	Dos ángulos que comparten un rayo. 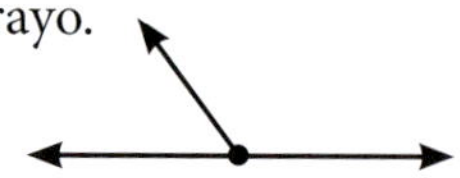
Algebraic Expression	An expression that contains numbers, operations and variables.	**Expresiones Algebraicas**	Una expresión que contiene números, operaciones y variables.
Alternate Exterior Angles	Two angles that are on the outside of two lines and are on opposite sides of a transversal. 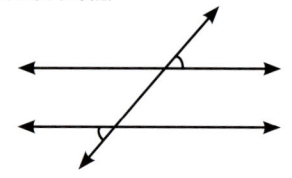	**Ángulos Exteriores Alternos**	Dos ángulos que están afuera de dos rectas y están a lados opuestos de una transversal. 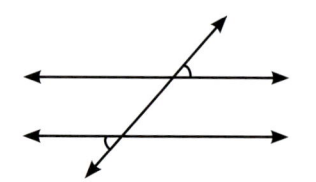
Alternate Interior Angles	Two angles that are on the inside of two lines and are on opposites sides of a transversal. 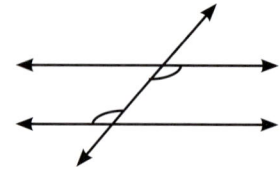	**Ángulos Interiores Alternos**	Dos ángulos que están adentro de dos rectas y están a lados opuestos de una transversal. 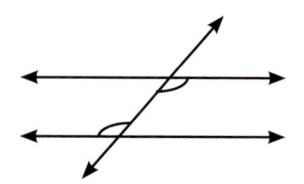
Angle	A figure formed by two rays with a common endpoint. 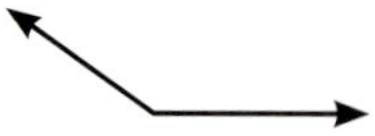	**Ángulo**	Una figura formada por dos rayos con un punto final en común. 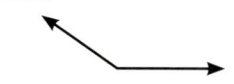

168 *Glossary ~ Glosario*

Area	The number of square units needed to cover a surface.	Área	El número de unidades cuadradas necesitadas para cubrir una superficie.
Ascending Order	Numbers arranged from least to greatest.	Progresión Ascendente	Los números ordenados de menor a mayor.
Associative Property	A property that states that numbers in addition or multiplication expressions can be grouped without affecting the value of the expression.	Propiedad Asociativa	Una propiedad que establece que los números en expresiones de suma o de multiplicación pueden ser agrupados sin afectar el valor de la expresión.
Axes	A horizontal and vertical number line on a coordinate plane.	Ejes	Una recta numérica horizontal y vertical en un plano de coordenadas.
Axis of Symmetry	The line of symmetry on a parabola that goes through the vertex.	El Eje De Las Simetría	La linia de simetría de una parábola que pasa por el vértice.

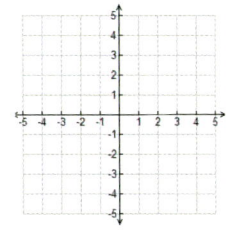

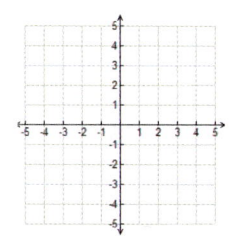

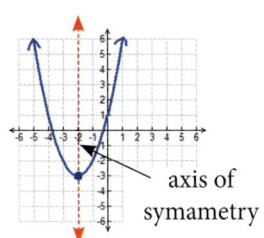

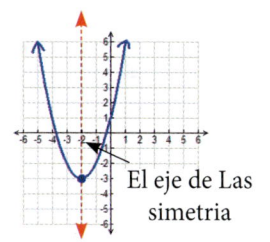

B

Bar Graph	A graph that uses bars to compare the quantities in a categorical data set.	Gráfico de Barras	Una gráfica que utiliza barras para comparar las cantidades en un conjunto de datos categórico.
Base (of a power)	The base of the power is the repeated factor. In x^a, x is the base.	Base (de la potencia)	La base de la potenciación es el factor repatidio. En x^a, x es la base.

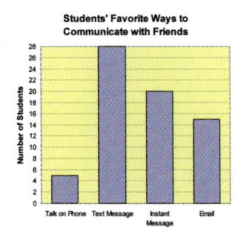

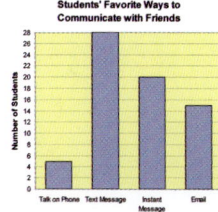

Glossary ~ Glosario **169**

Base (of a solid)	See Prism, Cylinder, Pyramid and Cone.	Base (de un sólido)	Ver Prisma, Cilindro, Pirámide y Cono.
Base (of a triangle)	Any side of a triangle.	Base (de un triángulo)	Cualquier lado de un triángulo.
Bias	A problem when gathering data that affects the results of the data.	Sesgo	Un problema que ocurre cuando se recogen datos que afectan los resultados de los datos.
Biased Sample	A group from a population that does not accurately represent the entire population.	Muestra Sesgada	Un grupo de una población que no representa con exactitud la población entera.
Binomials	Expressions involving two terms (i.e. $x - 2$).	Binomiales	Expresiones que impliquen dos terminos. (es decir: $x - 2$).
Bivariate Data	Data that describes two variables and looks at the relationship between the two variables.	Datos de dos Variables	Los datos que describen dos variables y analiza la relación entre estas dos variables.
Box-and-Whisker Plot	A diagram used to display the five-number summary of a data set.	Diagrama de Líneas y Bloques	Un diagrama utilizado para mostrar el resumen de cinco números de un conjunto de datos.

C

Categorical Data	Data collected in the form of words.	Datos Categóricos	Datos recopilados en la forma de palabras.
Center of a Circle	The point inside a circle that is the same distance from all points on the circle.	Centro de un Círculo	Un ángulo dentro de un círculo que está a la misma distancia de todos los puntos en el círculo.

Central Angle	An angle in a circle with its vertex at the center of a circle.	Ángulo Central	Un ángulo en un círculo con su vértice en el centro del círculo.
Chord	A line segment with endpoints on a circle.	Cuerda	Un segmento de la recta con puntos finales en el círculo.
Circle	The set of all points that are the same distance from a center point.	Círculo	El conjunto de todos los puntos que están a la misma distancia de un punto central.
Circumference	The distance around a circle.	Circunferencia	La distancia alrededor de un círculo.
Coefficient	The number multiplied by a variable in a term.	Coeficiente	El número multiplicado por una variable en un término.
Commutative Property	A property that states numbers can be added or multiplied in any order.	Propiedad Conmutativa	Una propiedad que establece que los números pueden ser sumados o multiplicados en cualquier orden.
Compatible Numbers	Numbers that are easy to mentally compute; used when estimating products and quotients.	Números Compatibles	Números que son fáciles de calcular mentalmente; utilizado cuando se estiman productos y cocientes.
Complementary Angles	Two angles whose sum is 90°.	Ángulos Complementarios	Dos ángulos cuya suma es de 90°.
Complements	Two probabilities whose sum is 1. Together they make up all the possible outcomes without repeating any outcomes.	Complementos	Dos probabilidades cuya suma es de 1. Juntos crean todos los posibles resultados sin repetir alguno.

Glossary ~ Glosario **171**

Completing the Square	The creation of a perfect square trinomial by adding a constant to an expression in the form $x^2 + bx$.	Terminado el Cuadrado	La creación de un trinomio cuadrado perfecto por adición de una constante a una expresión en la forma $x^2 + bx$.
Complex Fraction	A fraction that contains a fractional expression in its numerator, denominator or both. $$\frac{\frac{3}{4}}{\frac{3}{8}}$$	Fracción Compleja	Una fracción que contiene una expresión fraccionaria en su numerador, el denominador o ambos. $$\frac{\frac{3}{4}}{\frac{3}{8}}$$
Composite Figure	A geometric figure made of two or more geometric shapes.	Figura Compuesta	Una figura geométrica formada por dos o más formas geométricas.
Composite Number	A whole number larger than 1 that has more than two factors.	Número Compuesto	Un número entero mayor que el 1 con más de dos factores.
Composite Solid	A solid made of two or more three-dimensional geometric figures.	Sólido Compuesto	Un sólido formado por dos o más figuras geométricas tridimensionales.
Composition of Transformations	A series of transformations on a point.	Composición de Transformaciones	Una serie de transformaciones en un punto.
Compound Probability	The probability of two or more events occurring.	Compuesto de Probabilidad	La probabilidad de dos o más eventos ocurriendo.
Conditional Frequency	The ratio of the observed frequency to the total number of frequencies in a given category from an experiment or survey.	Frecuencia Condicional	La relación de una frecuencia observada para el número total de frecuencias en una categoría dada del experimento o encuesta.
Cone	A solid formed by one circular base and a vertex.	Cono	Un sólido formado por una base circular y una vértice.
Congruent	Equal in measure.	Congruente	Igual en medida.

Congruent Figures	Two shapes that have the exact same shape and the exact same size. 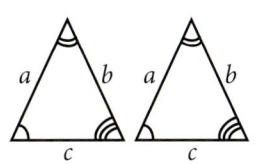	Figuras Congruentes	Dos figuras que tienen exactamente la misma forma y el mismo tamaño. 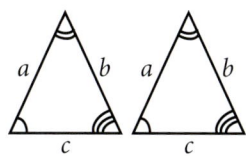
Constant	A term that has no variable.	Constante	Un término que no tiene variable.
Continuous	When a graph can be drawn from beginning to end without any breaks.	Continuo	Cuando una gráfica puede ser dibujada desde principio a fin sin ninguna interrupción.
Conversion	The process of renaming a measurement using different units.	Conversión	El proceso de renombrar una medida utilizando diferentes unidades.
Coordinate Plane	A plane created by two number lines intersecting at a 90° angle. 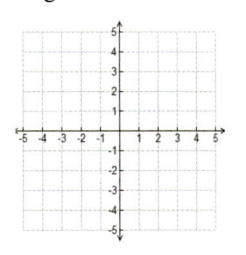	Plano de Coordenadas	Un plano creado por dos rectas numéricas que se intersecan a un ángulo de 90°. 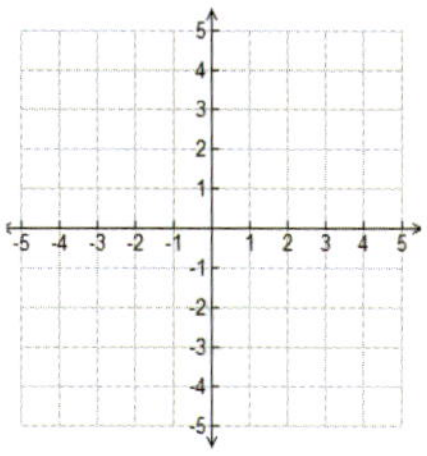
Correlation	The relationship between two variables in a scatter plot.	Correlación	La relación entre dos variables en un gráfico de dispersión.
Corresponding Angles	Two non-adjacent angles that are on the same side of a transversal with one angle inside the two lines and the other on the outside of the two lines. 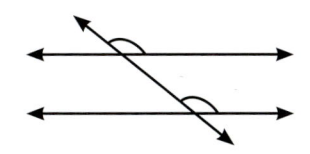	Ángulos Correspondientes	Dos ángulos no adyacentes que están en el mismo lado de una transversal con un ángulo adentro de las dos rectas y el otro afuera de las dos rectas. 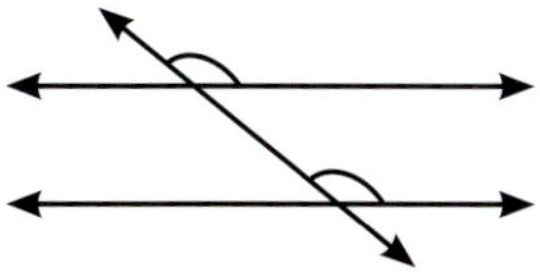
Corresponding Parts	The angles and sides in similar or congruent figures that match.	Partes Correspondientes	Los ángulos y lados en figuras similares o congruentes que concuerdan.

Glossary ~ Glosario **173**

Cube Root	One of the three equal factors of a number. $3 \cdot 3 \cdot 3 = 27 \quad \sqrt[3]{27} = 3$	Raíz Cúbica	Uno de los tres factores iguales de un número. $3 \cdot 3 \cdot 3 = 27 \quad \sqrt[3]{27} = 3$
Cubed	A term raised to the power of 3.	Cubicado	Un término elevado a la potencia de 3.
Cylinder	A solid formed by two congruent and parallel circular bases.	Cilindro	Un sólido formado por dos bases circulares congruentes y paralelas.

D

Decimal	A number with a digit in the tenths place, hundredths place, etc.	Decimal	Un número con un dígito en las décimas, las centenas, etc.
Degrees	A unit used to measure angles.	Grados	Una unidad utilizada para medir ángulos.
Dependent Events	Two (or more) events such that the outcome of one event affects the outcome of the other event(s).	Eventos Dependiente	Dos (o más) eventos de tal manera que el resultado de un evento afecta el resultado del otro evento (s).
Dependent Variable	The variable in a relationship that depends on the value of the independent variable.	Variable Dependiente	La variable en una relación que depende del valor de la variable independiente.
Descending Order	Numbers arranged from greatest to least.	Progresión Descendente	Los números ordenados de mayor a menor.

Diameter	The distance across a circle through the center. 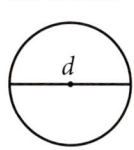	Diámetro	La distancia a través de un círculo por el centro. 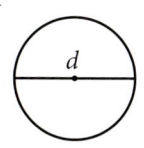
Dilation	A transformation which changes the size of the figure but not the shape. 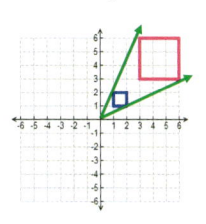	Dilatación	Una transformación que cambia el tamaño de la figura, pero no la forma. 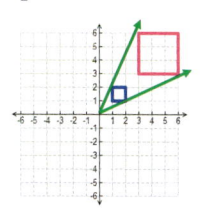
Direct Variation	A linear function that passes through the origin and has the equation $y = mx$.	Variación Directa	Una función lineal que pasa a través del origen y tiene la ecuación $y = mx$.
Discount	The decrease in the price of an item.	Descuento	La disminución de precio en un artículo.
Discrete	When a graph can be represented by a unique set of points rather than a continuous line.	Discreto	Cuando una gráfica puede ser representada por un conjunto de puntos único en vez de una recta continua.
Discriminant	In the quadratic formula, the expression under the radical sign. The discriminant provides information about the number of real roots or solutions of a quadratic equation. $$\frac{-b \pm \sqrt{b^2 - 4ac}}{2a}$$	Discriminante	En la fórmula cuadrática, la expresión bajo el signo radical. El discriminante proporciona información sobre el número o las verdaderas raíces o soluciones de una ecuación cuadrática. $$\frac{-b \pm \sqrt{b^2 - 4ac}}{2a}$$
Distance Formula	A formula used to find the distance between two points on the coordinate plane. $$d = \sqrt{(x_2 - x_1)^2 + (y_2 - y_1)^2}$$	Fórmula de Distancia	Una fórmula utilizada para encontrar la distancia entre dos puntos en un plano de coordenadas. $$d = \sqrt{(x_2 - x_1)^2 + (y_2 - y_1)^2}$$

Glossary ~ Glosario **175**

Distributive Property	A property that can be used to rewrite an expression without parentheses. $a(b + c) = ab + ac$	Propiedad Distributiva	Una propiedad que puede ser utilizada para reescribir una expresión sin paréntesis: $a(b + c) = ab + ac$
Dividend	The number being divided. **100** $\div 4 = 25$	Dividendo	El número que es dividido. **100** $\div 4 = 25$
Divisor	The number used to divide. $100 \div$ **4** $= 25$	Divisor	El número utilizado para dividir. $100 \div$ **4** $= 25$
Domain	The set of input values of a function.	El Dominio	El conjunto de valores entrados de la función.
Dot Plot	A data display that consists of a number line with dots equally spaced above data values.	Punto de Gráfico	Una visualización de datos que consiste de una línea numérica con puntos igualmente espaciados sobre valores de datos.
Double Stem-and-Leaf Plot	A stem-and-leaf plot where one set of data is placed on the right side of the stem and another is placed on the left of the stem.	Doble Gráfica de Tallo y Hoja	Una gráfica de tallo y hoja donde un conjunto de datos es colocado al lado derecho del tallo y el otro es colocado a la izquierda del tallo.

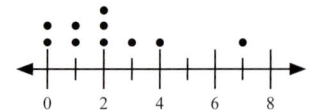

E

Edge	The segment where two faces of a solid meet.	Arista (Borde)	El segmento donde dos caras de un sólido se encuentran.

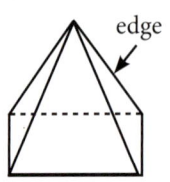

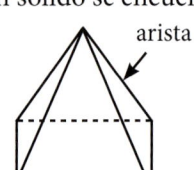

176 *Glossary ~ Glosario*

English	Definition	Spanish	Definición
Elimination Method	A method for solving a system of linear equations.	Método de Eliminación	Un método para resolver un sistema de ecuaciones lineales.
Enlargement	A dilation that creates an image larger than its pre-image.	Agrandamiento	Una dilatación que crea una imagen más grande que su pre-imagen.
Equally Likely	Two or more possible outcomes of a given situation that have the same probability.	Igualmente Probables	Dos o más posibles resultados de una situación dada que tienen la misma probabilidad.
Equation	A mathematical sentence that contains an equals sign between 2 expressions.	Ecuación	Una oración matemática que contiene un símbolo de igualdad entre dos expresiones.
Equiangular	A polygon in which all angles are congruent.	Equiángulo	Un polígono en el cual todos los ángulos son congruentes.
Equilateral	A polygon in which all sides are congruent.	Equilátero	Un polígono en el cual todos los lados son congruentes.
Equivalent Decimals	Two or more decimals that represent the same number.	Decimales Equivalentes	Dos o más decimales que representan el mismo número.
Equivalent Expressions	Two or more expressions that represent the same algebraic expression.	Expresiones Equivalentes	Dos o más expresiones que representan la misma expresión algebraica.
Equivalent Fractions	Two or more fractions that represent the same number.	Fracciones Equivalentes	Dos o más fracciones que representan el mismo número.
Evaluate	To find the value of an expression.	Evaluar	Encontrar el valor de una expresión.
Even Distribution	A set of data values that is evenly spread across the range of the data.	Distribución Igualada	Un conjunto de valores de datos que es esparcido de modo uniforme a través del rango de los datos.

Event	A desired outcome or group of outcomes.	Evento	Un resultado o grupo de resultados deseados.
Experimental Probability	The ratio of the number of times an event occurs to the total number of trials.	Probabilidad Experimental	La razón de la cantidad de veces que un suceso ocurre a la cantidad total de intentos.
Exponent	In x^a, a is the exponent. The exponent shows the number of times the factor (x) is repeated.	Exponente	En x^a, a es el exponente. El exponente indica el número de veces que se repite el factor (x).
Exponential Function	A function that can be described by an equation in the form $f(x) = bm^x$.	Función Exponencial	Una función que puede ser descrito por una ecuación en la forma $f(x) = bm^x$.

F

Face	A polygon that is a side or base of a solid.	Cara	Un polígono que es una base de lado de un sólido.
Factors	Whole numbers that can be multiplied together to find a product.	Factores	Números enteros que pueden ser multiplicados entre si para encontrar un producto.
First Quartile (Q1)	The median of the lower half of a data set.	Primer Cuartil (Q1)	Mediana de la parte inferior de un conjunto de datos.
Five-Number Summary	Describes the spread of a data set using the minimum, 1st quartile, median, 3rd quartile and maximum.	Sumario de Cinco Números	Describe la extensión de un conjunto de datos utilizando el mínimo, el primer cuartil, la mediana el tercer cuartil y el máximo.
Formula	An algebraic equation that shows the relationship among specific quantities.	Fórmula	Una ecuación algebraica que enseña la relación entre cantidades específicas.
Fraction	A number that represents a part of a whole number, written as $\frac{numerator}{denominator}$.	Fracción	Un número que representa una parte de un número entero, escrito como $\frac{numerador}{denominador}$.

Frequency	The number of times an item occurs in a data set.	Frecuencia	La cantidad de veces que un artículo ocurre en un conjunto de datos.
Frequency Table	A table which shows how many times a value occurs in a given interval.	Tabla de Frecuencia	Una tabla que enseña cuantas veces un valor ocurre en un intervalo dado.
Function	A relationship between two variables that has one output value for each input value.	Función	Una relación entre dos variables que tiene un valor de salida para cada valor de entrada.

G

General Form	A quadratic function is in general form when written $f(x) = ax^2 + bx + c$ where $a \neq 0$.	Forma General	Una función cuadrática es en forma general cuándo escrito $f(x) = ax^2 + bx + c$ donde $a \neq 0$.
Geometric Probability	Ratios of lengths or areas used to find the likelihood of an event.	Probabilidad Geométrica	Razones de longitudes o áreas utilizadas para encontrar la probabilidad de un suceso.
Geometric Sequence	A list of numbers that begins with a starting value. Each term in the sequence is generated by multiplying the previous term in the sequence by a constant multiplier.	Secuenciación Geométrica	Una lista de números que comienza con un valor inicial. Cada término de la secuencia se genera al multiplicar el término anterior de la secuencia por un multiplicar constante.
Greatest Common Factor (GCF)	The greatest factor that is common to two or more numbers.	Máximo Común Divisor (MCD)	El máximo divisor que le es común a dos o más números.
Grouping Symbols	Symbols such as parentheses or fraction bars that group parts of an expression.	Símbolos de Agrupación	Símbolos como el paréntesis o barras de fracción que agrupan las partes de una expresión.

H

Height of a Triangle	A perpendicular line drawn from the side whose length is the base to the opposite vertex.	Altura de un Triángulo	Una recta perpendicular dibujada desde el lado cuya longitud es la base al vértice opuesto.
Histogram	A bar graph that displays the frequency of numerical data in equal-sized intervals.	Histograma	Un gráfico de barras que muestra la frecuencia de datos numéricos en intervalos de tamaños iguales.
Hypotenuse	The side opposite the right angle in a right triangle.	Hipotenusa	El lado opuesto el ángulo recto en un triángulo rectángulo.

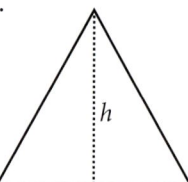

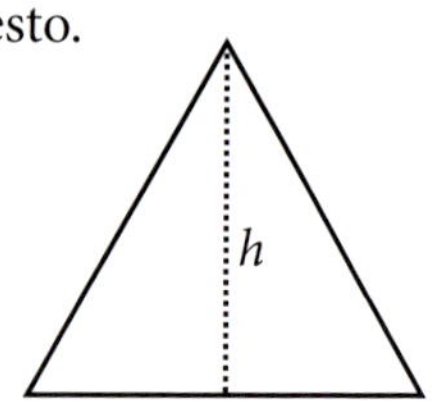

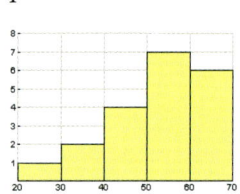

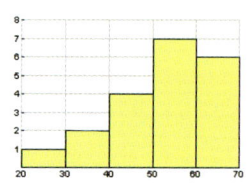

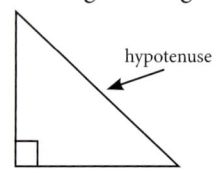

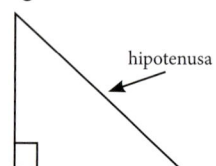

I-J-K

Image	A point or figure which is the result of a transformation or series of transformations.	Imagen	Un punto o figura que es el resultado de una transformación o una serie de transformaciones.
Improper Fraction	A fraction whose numerator is greater than or equal to its denominator.	Fracción Impropia	Una fracción cuyo numerador es mayor o igual a su denominador.
Independent Events	Two (or more) events such that the outcome of one event does not affect the outcome of the other event(s).	Eventos Independientes	Dos (o más) eventos de tal manera que el resultado de un evento no afecta el resultado del otro evento (s).

180 Glossary ~ Glosario

English	Definition	Spanish	Definición
Independent Variable	The variable representing the input values.	Variable Independiente	La variable que representa los valores entratos.
Inequality	A mathematical sentence using <, >, ≤ or ≥ to compare two quantities.	Desigualdad	Un enunciado matemático usando <, >, ≤ ó ≥ para comparar dos cantidades.
Inference	A logical conclusion based on known information.	Inferencia	Una conclusión lógica basada en la información conocida.
Input-Output Table	A table used to describe a function by listing input values with their output values.	Tabla de Entrada y Salida	Una tabla utilizada para describir una función al enumerar valores de entrada con sus valores de salidas.
Integers	The set of all whole numbers, their opposites, and 0.	Enteros	El conjunto de todos los números enteros, sus opuestos y 0.
Interquartile Range (IQR)	The difference between the 3rd quartile and the 1st quartile in a set of data.	Rango Intercuartil (IQR)	La diferencia entre el tercer cuartil y el primer cuartil en un conjunto de datos.
Inverse Operations	Operations that undo each other.	Operaciones Inversas	Operaciones que se cancelan la una a la otra.
IQR Method	A method for determining outliers using interquartile ranges.	Método IQR	Un método para determinar los datos aberrantes.
Irrational Numbers	A number that cannot be expressed as a fraction of two integers.	Números Irracionales	Un número que no puede ser expresado como una fracción de dos enteros.

Isosceles Trapezoid	A trapezoid that has congruent legs.	Trapezoide Isósceles	Un trapezoide con catetos congruentes.
Isosceles Triangle	A triangle that has two or more congruent sides.	Triángulo Isósceles	Un triángulo que tiene dos o más lados congruentes.

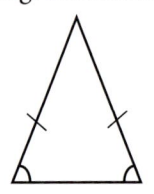

L

Lateral Face	A side of a solid that is not a base.	Cara Lateral	Un lado de un sólido que no sea una base.
Least Common Denominator (LCD)	The least common multiple of two or more denominators.	Mínimo Común Denominador (MCD)	El mínimo común múltiplo de dos o más denominadores.
Least Common Multiple (LCM)	The smallest nonzero multiple that is common to two or more numbers.	Mínimo Común Múltiplo (MCM)	El múltiplo más pequeño que no sea cero que le es común a dos o más números.
Leg	The two sides of a right triangle that form a right angle.	Cateto	Los dos lados de un triángulo rectángulo que forman un ángulo recto.
Like Terms	Terms that have the same variable(s).	Términos Semejantes	Términos que tienen el mismo variable(s).

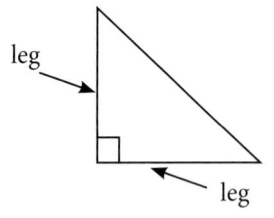

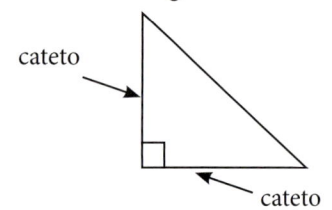

182 *Glossary ~ Glosario*

Line of Best Fit	A line which best represents the pattern of a two-variable data set.	Recta de Mejor Ajuste	Una recta que mejor representa el patrón de un conjunto de datos de dos variables.
Linear Equation	An equation whose graph is a line.	Ecuación Lineal	Una ecuación cuya gráfica es una recta.
Linear Function	A function whose graph is a line.	Función Lineal	Una función cuya gráfica es una recta.
Linear Pair	Two adjacent angles whose non-common sides are opposite rays.	Par Lineal	Dos ángulos adyacentes cuyos lados no comunes son rayos opuestos.

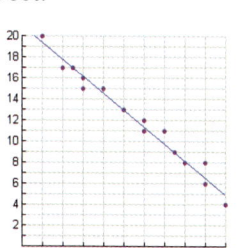

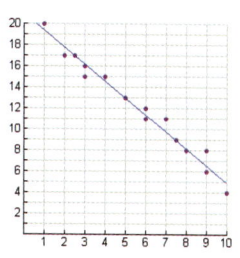

M

Mark-up	The increase in the price of an item.	Margen de Beneficio	El aumento de precio en un artículo.
Maximum	The highest point on a curve.	Máximo	El punto más alto en la curva.
Mean	The sum of all values in a data set divided by the number of values.	Media	La suma de todos los valores en un conjunto de datos dividido entre la cantidad de valores.
Mean Absolute Deviation	A statistic that shows the average distance from the mean for all numbers in a data set.	Desviación Media Absoluta	Una estadística que muestra la distancia promedio entre la media de todos los números en una serie de datos.

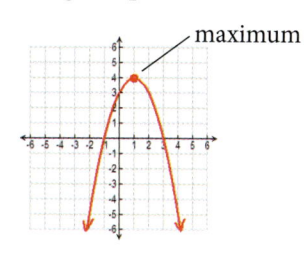

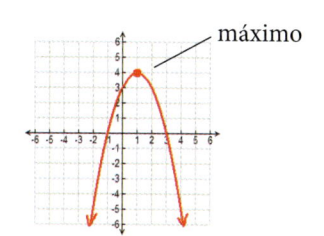

Glossary ~ Glosario

Measures of Center	Numbers that are used to represent a data set with a single value; the mean, median, and mode are the measures of center.	Medidas de Centro	Números que son utilizados para representar un conjunto de datos con un solo valor; la media, la mediana, y la moda son las medidas de centro.
Measures of Variability	Statistics that help determine the spread of numbers in a data set.	Medidas de Variabilidad	Las estadísticas que ayudan a determinar la extensión de los números en una serie de datos.
Median	The middle number or the average of the two middle numbers in an ordered data set.	Mediana	El número medio o el promedio de los dos números medios en un conjunto de datos ordenados.
Minimum	The lowest point on a curve.	Mínimo	El punto más bajo en la curva.
Mixed Number	The sum of a whole number and a fraction less than 1.	Números Mixtos	La suma de un número entero y una fracción menor que 1.
Mode	The number(s) or item(s) that occur most often in a data set.	Moda	El número(s) o artículo(s) que ocurre con más frecuencia en un conjunto de datos.
Motion Rate	A rate that compares distance to time.	Índice de Movimiento	Un índice que compara distancia por tiempo.
Multiple	The product of a number and nonzero whole number.	Múltiplo	El producto de un número y un número entero que no sea cero.

N

Negative Number	A number less than 0.	Número Negativo	Un número menor que 0.

English	Definition	Spanish	Definición
Net	A two-dimensional pattern that folds to form a solid.	Red	Un patrón bidimensional que se dobla para formar un sólido.
Non-Linear Function	A function whose graph does not form a line.	Ecuación No Lineal	Una ecuación cuya gráfica no forma una recta.
Normal Distribution	A set of data values where the majority of the values are located in the middle of the data set and can be displayed by a bell-shaped curve.	Distribución Normal	Un conjunto de valores de datos donde la mayoría de los valores están localizados en el medio del conjunto de datos y pueden ser mostrados por una curva de forma de campana.
Numerical Data	Data collected in the form of numbers.	Datos Numéricos	Datos recopilados en la forma de números.
Numerical Expressions	An expression consisting of numbers and operations that represents a specific value.	Expresiones Numéricas	Una expresión que consta de números y operaciones que representa un valor específico.

O

English	Definition	Spanish	Definición
Obtuse Angle	An angle that measures more than 90° but less than 180°.	Ángulo Obtuso	Un ángulo que mide más de 90° pero menos de 180°.
Opposites	Numbers that are the same distance from 0 on a number line but are on opposite sides of 0.	Opuestos	Números a la misma distancia del 0 en un recta numérica pero en lados opuestos del 0.
Order of Operations	The rules to follow when evaluating an expression with more than one operation.	Orden de Operaciones	Las reglas a seguir cuando se evalúa una expresión con más de una operación.
Ordered Pair	A pair of numbers used to locate a point on a coordinate plane (x, y).	Par Ordenado	Un par de números utilizados para localizar un punto en un plano de coordenadas (x, y).

Origin	The point where the *x*- and *y*-axis intersect on a coordinate plane (0, 0).	Origen	El punto donde el eje de la *x*- *y* el de la *y*- se cruzan en un plano de coordenadas (0,0).
Outcome	One possible result from an experiment or a probability sample space.	Resultado	Un resultado posible de un experimento o un espacio de probabilidad de la muestra.
Outlier	An extreme value that varies greatly from the other values in a data set.	Dato Aberrante	Un valor extremo que varía mucho de los otros valores en un conjunto de datos.

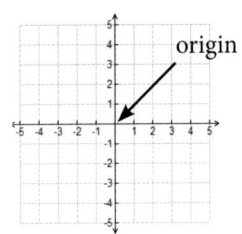

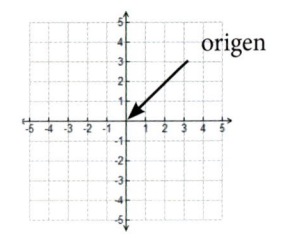

P

Parabola	The graph of a quadratic function.	Parábola	La gráfica de una función cuadratica.
Parallel	Lines in the same plane that never intersect.	Paralela	Rectas en el mismo plano que nunca se intersecan.
Parallel Box-and-Whisker Plot	One box-and-whisker plot placed above another; often used to compare data sets.	Diagrama Paralelo de Líneas y Bloques	Un diagrama de líneas y bloques ubicado sobre otro para comparar conjuntos de datos.

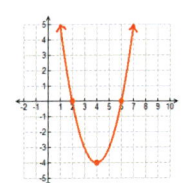

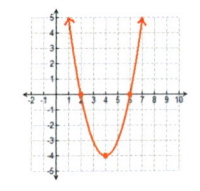

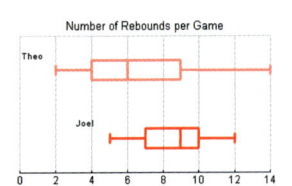

186 *Glossary ~ Glosario*

Parallelogram	A quadrilateral with both pairs of opposite sides parallel.	Paralelogramo	Un cuadrilateral con ambos pares de lados opuestos paralelos.
Parent Function	The simplest form of a particular type of function.	Función Principal	La forma más simple de un tipo particular de la función.
Parent Graph	The most basic graph of a function.	Gráfico Matriz	La gráfica más básica de una función.
Percent	A ratio that compares a number to 100.	Por Ciento	Una razón que compara un número con 100.
Percent of Change	The percent a quantity increases or decreases compared to the original amount.	Por Ciento de Cambio	El por ciento que una cantidad aumenta o disminuye comparado a la cantidad original.
Percent of Decrease	The percent of change when the new amount is less than the original amount.	Por Ciento de Disminución	El por ciento de cambio cuando la nueva cantidad es menos que la cantidad original.
Percent of Increase	The percent of change when the new amount is more than the original amount.	Por Ciento de Incremento	El por ciento de cambio cuando la nueva cantidad es más que la cantidad original.
Perfect Cube	A number whose cube root is an integer.	Cubo Perfecto	Un número cuyo raíz cúbica es un número entero.
Perfect Square	A number whose square root is an integer.	Cuadrado Perfecto	Un número cuyo raíz cuadrado es un número entero.
Perfect Square Trinomial	A trinomial that is the square of a binomial.	Trinomio Cuadrado Perfecto	Un trinomio que es el cuadrado de un binomio.
Perimeter	The distance around a figure.	Perímetro	La distancia alrededor de una figura.

Perpendicular	Two lines or segments that form a right angle.	Perpendicular	Dos rectas o segmentos que forman un ángulo recto.
Pi (π)	The ratio of the circumference of a circle to its diameter.	Pi (π)	La razón de la circunferencia de un círculo a su diámetro.
Pictograph	A graph that uses pictures to compare the amounts represented in a categorical data set.	Gráfica Pictórica	Una gráfica que utiliza dibujos para comparar las cantidades representadas en un conjunto de datos categóricos.
Pie Chart	A circle graph that shows information as sectors of a circle.	Gráfico Circular	Enseña la información como sectores de un círculo.
Polygon	A closed figure formed by three or more line segments.	Polígono	Una figura cerrada formada por tres o más segmentos de rectas.
Population	The entire group of people or objects one wants to gather information about.	Población	Todo el grupo de personas o los objetos a los que se quiere obtener información sobre.
Positive Number	A number greater than 0.	Número Positivo	Un número mayor que 0.
Power	An expression such as x^a which consists of two parts, the base (x) and the exponent (a).	Potencia	Una expresión como x^a que consiste de dos partes, la base (x) y el exponente (a).
Pre-image	The original figure prior to a transformation.	Pre-imagen	La figura original antes de una transformación.

Prime Factorization	When any composite number is written as the product of all its prime factors.	Factorización Prima	Cuando cualquier número compuesto es escrito como el producto de todos los factores primos.
Prime Number	A whole number larger than 1 that has only two possible factors, 1 and itself.	Número Primo	Un número entero mayor que 1 que tiene solo dos factores posibles, 1 y el mismo.
Prism	A solid formed by polygons with two congruent, parallel bases.	Prisma	Un sólido formado por polígonos con dos bases congruentes y paralelas.
Probability	The measure of how likely it is an event will occur.	Probabilidad	La medida de cuán probable un suceso puede ocurrir.
Product	The answer to a multiplication problem.	Producto	La respuesta a un problema de multiplicación.
Proper Fraction	A fraction with a numerator that is less than the denominator.	Fracción Propia	Una fracción con un numerador que es menos que el denominador.
Proportion	An equation stating two ratios are equivalent.	Proporción	Una ecuación que establece que dos razones son equivalentes.
Protractor	A tool used to measure angles.	Transportador	Una herramienta para medir ángulos.
Pyramid	A solid with a polygonal base and triangular sides that meet at a vertex.	Pirámide	Un sólido con una base poligonal y lados triangulares que se encuentran en un vértice.

Glossary ~ Glosario **189**

Pythagorean Triple	A set of three positive integers (a, b, c) such that $a^2 + b^2 = c^2$.	Triple de Pitágoras	Un conjunto de tres enteros positivos (a, b, c) tal que $a^2 + b^2 = c^2$.

Q

Q-Points	Points that are created by the intersection of the quartiles for the *x*- and *y*-values of a two-variable data set.	Puntos Q	Puntos que son creados por la intersección de los cuartiles para los valores de la *x*- y la *y*- de un conjunto de datos de dos variables.
Quadrants	Four regions formed by the *x* and *y* axes on a coordinate plane.	Cuadrantes	Cuatro regiones formadas por el eje-*x* y el eje-*y* en un plano de coordenadas.
Quadratic Formula	A method which can be used to solve quadratic equations in the form $0 = ax^2 + bx + c$, where $a \neq 0$. $$x = \frac{-b \pm \sqrt{b^2 - 4ac}}{2a}$$	Fórmula Cuadrática	Un método que puede usarse para resolver ecuaciones cuadraticas en la forma $0 = ax^2 + bx + c$, donde $a \neq 0$. $$x = \frac{-b \pm \sqrt{b^2 - 4ac}}{2a}$$
Quadratic Function	Any function in the family with the parent function of $f(x) = x^2$.	Función Cuadrática	Cualquier otra función en la familia con la función principal de $f(x) = x^2$.
Quadrilateral	A polygon with four sides.	Cuadrilateral	Un polígono con cuatro lados.
Quotient	The answer to a division problem.	Cociente	La solución a un problema de división.

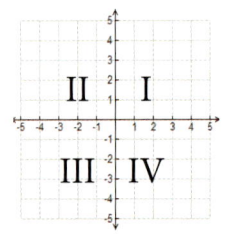

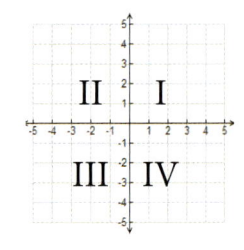

R

Radius	The distance from the center of a circle to any point on the circle.	Radio	La distancia desde el centro de un círculo a cualquier punto en el círculo.

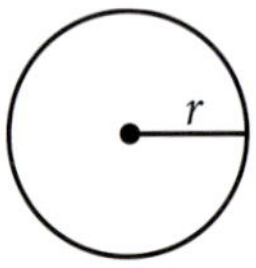

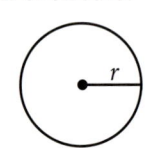

Glossary ~ Glosario

Random Sample	A group from a population created when each member of the population is equally likely to be chosen.	Muestra Aleatoria	Un grupo de una población creada cuando cada miembro de la población tiene la misma probabilidad de ser elegido.
Range (of a data set)	The difference between the maximum and minimum values in a data set.	Rango	La diferencia entre los valores máximo y mínimo en un conjunto de datos.
Range (of a function)	The set of output values of a function.	Rango (de una función)	El conjunto de valores salidos de la función.
Rate	A ratio of two numbers that have different units.	Índice	Una proporción de dos números con diferentes unidades.
Rate Conversion	A process of changing at least one unit of measurement in a rate to a different unit of measurement.	Conversión de Índice	Un proceso de cambiar por lo menos una unidad de medición en un índice a una diferente unidad de medición.
Rate of Change	The change in y-values over the change in x-values on a linear graph.	Índice de Cambio	El cambio en los valores de y sobre el cambio en los valores de x en una gráfica lineal.
Ratio	A comparison of two numbers using division. $a:b \quad \dfrac{a}{b} \quad a \text{ to } b$	Razón	Una comparación de dos números utilizando división. $a:b \quad \dfrac{a}{b} \quad a \text{ a } b$
Rational Number	A number that can be expressed as a fraction of two integers.	Número Racional	Un número que puede ser expresado como una fracción de dos enteros.
Ray	A part of a line that has one endpoint and extends forever in one direction.	Rayo	Una parte de una recta que tiene un punto final y se extiende eternamente en una dirección.
Real Numbers	The set of numbers that includes all rational and irrational numbers.	Números Racionales	El conjunto de números que incluye todos los números racionales e irracionales.

Glossary ~ Glosario **191**

Reciprocals	Two numbers whose product is 1.	Recíprocos	Dos números cuyo producto es 1.
Rectangle	A quadrilateral with four right angles.	Rectángulo	Un cuadrilátero con cuatro ángulos rectos.
Recursive Routine	A routine described by stating the start value and the operation performed to get the following terms.	Rutina Recursiva	Una rutina descrita al exponer el valor del comienzo y la operación realizada para conseguir los términos siguientes.
Recursive Sequence	An ordered list of numbers created by a first term and a repeated operation.	Secuencia Recursiva	Una lista de números ordenados creada por un primer término y una operación repetida.
Reduction	A dilation that creates an image smaller than its pre-image.	Reducción	Una dilatación que crea una imagen más pequeña que su pre-imagen.
Reflection	A transformation in which a mirror image is produced by flipping a figure over a line.	Reflexión	Una transformación en el que se produce una imagen reflejada volteando una figura sobre una línea.
Relative Frequency	The ratio of the observed frequency to the total number of frequencies in an experiment or survey.	Frecuencia Relativa	La proporción de la frecuencia observada para el número total de frecuencias en un experimento o estudio.
Remainder	A number that is left over when a division problem is completed.	Remanente	Un número que queda cuando un problema de división se ha completado.
Repeating Decimal	A decimal that has one or more digits that repeat forever.	Decimal Repetitivo	Un decimal que tiene uno o más dígitos que se repiten eternamente.

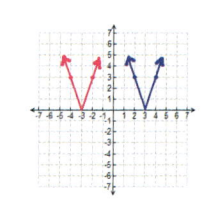

Representative Sample	A group from a population that accurately represents the entire population.	Muestra Representativa	Un grupo de una población que representa con precisión toda la población.
Rhombus	A quadrilateral with four sides equal in measure.	Rombo	Un cuadrilátero con cuatro lados iguales en la medida.
Right Angle	An angle that measures 90°.	Ángulo Recto	Un ángulo que mide 90°.
Roots	The *x*-intercepts of a quadratic function. (zeros, roots, *x*-intercepts)	Raíces	Las intersecciones-*x* de una función cuadratica. (ceros, raíces, intersecciones-*x*)
Rotation	A transformation which turns a point or figure about a fixed point, often the origin.	Rotación	Una transformación que convierte un punto a una figura sobre un punto fijo.

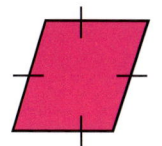

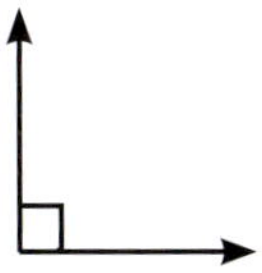

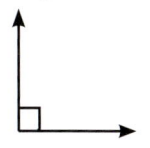

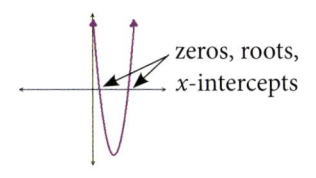

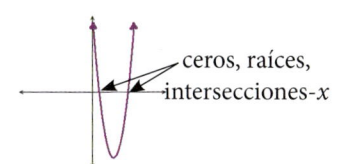

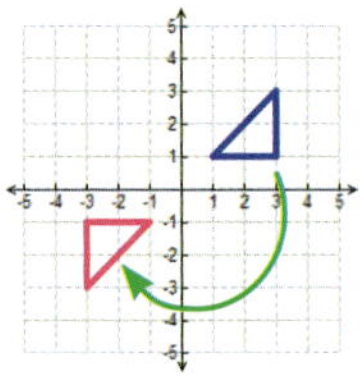

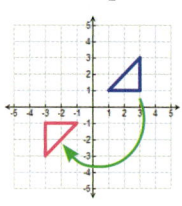

S

Sales Tax	An amount added to the cost of an item. The amount added is a percent of the original amount as determined by a state, county or city.	Impuesto sobre las Ventas	Una cantidad añadida al costo de un artículo. La cantidad añadida es un por ciento de la cantidad original determinado por el estado, condado o ciudad.

Glossary ~ Glosario **193**

Same-Side Interior Angles	Two angles that are on the inside of two lines and are on the same side of a transversal.	Ángulos Interiores del Mismo Lado	Dos ángulos que están en el interior de dos rectas y están en el mismo lado de una transversal.
Sample	A group from a population that is used to make conclusions about the entire population.	Muestra	Un grupo de una población que se utiliza para sacar conclusiones sobre toda la población.
Sample Space	The set of all possible outcomes.	Muestra de Espacio	El conjunto de todos los resultados posibles.
Scale	The ratio of a length on a map or model to the actual object.	Escala	La razón de una longitud en un mapa o modelo al objeto verdadero.
Scale Factor	The ratio of corresponding sides in two similar figures.	Factor de Escala	La razón de los lados correspondientes en dos figuras similares.
Scalene Triangle	A triangle that has no congruent sides.	Triángulo Escaleno	Un triángulo sin lados congruentes.
Scatter Plot	A set of ordered pairs graphed on a coordinate plane.	Diagrama de Dispersión	Un conjunto de pares ordenados graficados en un plano de coordenadas.
Scientific Notation	Scientific notation is an exponential expression using a power of 10 where $1 \leq N < 10$ and P is an integer. $N \times 10^P$	La Notación Científica	Notación científica es una expresión exponencial con una potencia de 10, donde $1 \leq N < 10$ y P es un número entero. $N \times 10^P$

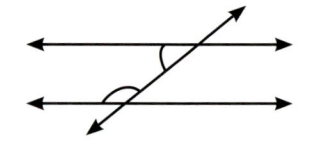

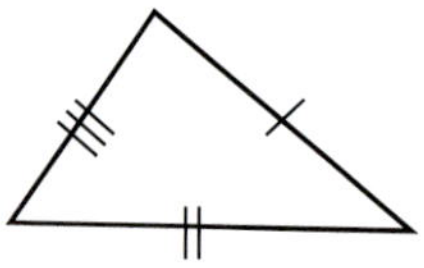

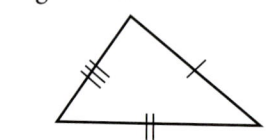

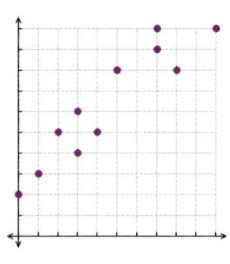

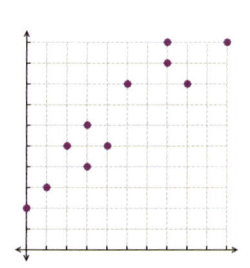

Sector	A portion of a circle enclosed by two radii.	Sector	Una porción de un circulo encerado por dos radios.
Sequence	An ordered list of numbers.	Sucesión	Una lista de números ordenados.
Similar Figures	Two figures that have the exact same shape, but not necessarily the exact same size.	Figuras Similares	Dos figuras que tienen exactamente la misma forma, pero no necesariamente el mismo tamaño exacto.
Similar Solids	Solids that have the same shape and all corresponding dimensions are proportional.	Sólidos Similares	Sólidos con la misma forma y todas sus dimensiones correspondientes son proporcionales.
Simplest Form	A fraction whose numerator and denominator's only common factor is 1.	Expresión Simple	Una fracción cuyo único factor común del numerador y del denominador es 1.
Simplify an Expression	To rewrite an expression without parentheses and combine all like terms.	Simplificar una Expresión	Reescribir una expresión sin paréntesis y combinar todos los términos iguales.
Simulation	An experiment used to model a situation.	Simulación	Un experimento utilizado para modelar una situación.
Single-Variable Data	A data set with only one type of data.	Datos de una Variable	Un conjunto de datos con tan solo un tipo de datos.
Sketch	To make a figure free hand without the use of measurement tools.	Esbozo	Hacer una figura a mano libre sin utilizar herramientas de medidas.
Skewed Left	A plot or graph with a longer tail on the left-hand side.	Torcido a la Izquierda	Un gráfico con una cola al lado izquierdo.

Glossary ~ Glosario **195**

Skewed Right	A plot or graph with a longer tail on the right-hand side.	Torcido a la Derecha	Un gráfico con una cola al lado derecho.
Slant Height	The height of a lateral face of a pyramid or cone.	Altura Sesgada	La altura de un cara lateral de una pirámide o cono.
Slope	The ratio of the vertical change to the horizontal change in a linear graph.	Pendiente	La razón del cambio vertical al cambio horizontal en una gráfica lineal.
Slope Triangle	A right triangle formed where one leg represents the vertical rise and the other leg is the horizontal run in a linear graph.	Triángulo de Pendiente	Un triángulo rectángulo formado donde una cateto representa el ascenso y la otra es una carrera horizontal en una gráfica lineal.
Slope-Intercept Form	A linear equation written in the form $y = mx + b$.	Forma de las Intersecciones con la Pendiente	Una ecuación lineal escrita en la forma $y = mx + b$.
Solid	A three-dimensional figure that encloses a part of space.	Sólido	Una figura tridimensional que encierra una parte del espacio.
Solution	Any value or values that makes an equation true.	Solución	Cualquier valor o valores que hacen una ecuación verdadera.
Solution of a System of Linear Equations	The ordered pair that satisfies both linear equations in the system.	Solución de un Sistema de Ecuaciones Lineales	El par ordenado que satisface ambas ecuaciones lineales en el sistema.

Sphere	A solid formed by a set of points in space that are the same distance from a center point.	Esfera	Un sólido formado por un conjunto de puntos en el espacio que están a la misma distancia de un punto central.
Square	A quadrilateral with four right angles and four congruent sides.	Cuadrado	Un cuadrilátero con cuatro ángulos rectos y cuatro lados congruente.
Square Root	One of the two equal factors of a number. $$3 \cdot 3 = 9 \quad 3 = \sqrt{9}$$	Raíz Cuadrada	Uno de los dos factores iguales de un número. $$3 \cdot 3 = 9 \quad 3 = \sqrt{9}$$
Squared	A term raised to the power of 2.	Cuadrado	Un término elevado a la potencia de 2.
Start Value	The output value that is paired with an input value of 0 in an input-output table.	Valor de Comienzo	El valor de salida que es aparejado con un valor de entrada de 0 en una tabla de entradas y salidas.
Statistics	The process of collecting, displaying and analyzing a set of data.	Estadísticas	El proceso de recopilar, exponer y analizar un conjunto de datos.
Stem-and-Leaf Plot	A plot which uses the digits of the data values to show the shape and distribution of the data set.	Gráfica de Tallo y Hoja	Un diagrama que utiliza los dígitos de los valores de datos para mostrar la forma y la distribución del conjunto de datos.
Straight Angle	An angle that measures 180°.	Ángulo Recto	Un ángulo que mide 180°.

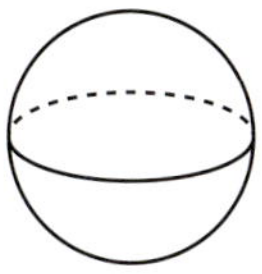

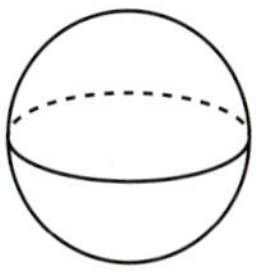

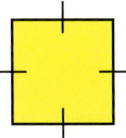

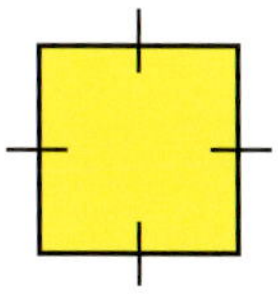

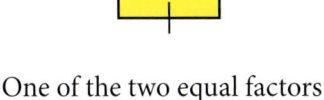

Glossary ~ Glosario

English	Definition	Spanish	Definición
Straight Edge	A ruler-like tool with no markings.	Borde Recto	Un gobernante como herramienta sin marcas.
Substitution Method	A method for solving a system of linear equations.	Método de Substitución	Un método para resolver un sistema de ecuaciones lineales.
Supplementary Angles	Two angles whose sum is 180°.	Ángulos Suplementarios	Dos ángulos cuya suma es 180°.
Surface Area	The sum of the areas of all the surfaces on a solid.	Área de la Superficie	La suma de las áreas de todas las superficies en un sólido.
System of Linear Equations	Two or more linear equations.	Sistema de Ecuaciones Lineales	Dos o más ecuaciones lineales.

T

English	Definition	Spanish	Definición
Term	A number or the product of a number and a variable in an algebraic expression; a number in a sequence.	Término	Un número o el producto de un número y una variable en una expresión algebraica; un número en una sucesión.
Terminating Decimal	A decimal that stops.	Decimal Terminado	Un decimal que para.
Theorem	A relationship in mathematics that has been proven.	Teorema	Una relación en las matemáticas que ha sido probada.
Theoretical Probability	The ratio of favorable outcomes to the number of possible outcomes.	Probabilidad Teórica	La proporción de resultados favorables a la cantidad de resultados posibles.
Third Quartile (Q3)	The median of the upper half of a data set.	Tercer Cuartil (Q3)	Mediana de la parte superior de un conjunto de datos.
Tick Marks	Equally divided spaces marked with a small line between every inch or centimeter on a ruler.	Marcas de Graduación	Espacios divididos igualmente marcados con una línea pequeña entre cada pulgada o centímetro en una regla.
Transformation	The movement of a figure on a graph so that it changes size or position.	Transformación	El movimiento de una figura en un gráfico de modo que cambia el tamaño o posición

Translation	A transformation in which a figure is shifted up, down, left or right.	Traducción	Una transformación donde la figura se mudo arriba, abajo, a la izquierda o a la derecha.
Transversal	A line that intersects two or more lines in the same plane.	Transversal	Una recta que interseca dos o más rectas en el mismo plano.
Trapezoid	A quadrilateral with exactly one pair of parallel sides.	Trapezoide	Un cuadrilateral con exactamente un par de lados paralelos.
Tree Diagram	A display that organizes information to determine possible outcomes.	Diagrama de Árbol	Una pantalla que organiza la información para determinar los posibles resulatados.
Trial	A single act of performing an experiment.	Prueba	Un solo intento de realizar un experimento.
Trinomial	An expression with three terms (i.e. $x^2 - 3x + 4$).	Trinomio	Una expreción que tiene tres terminos (es decir: $x^2 - 3x + 4$).
Two-Step Equation	An equation that has two different operations.	Ecuación de Dos Pasos	Una ecuación que tiene dos operaciones diferentes.
Two-Variable Data	A data set where two groups of numbers are looked at simultaneously.	Datos de dos Variables	Un conjunto de datos dónde dos grupos de números se observan simultáneamente.

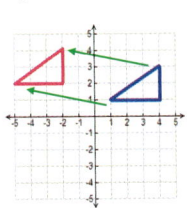

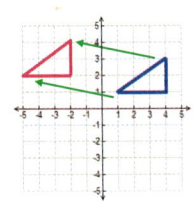

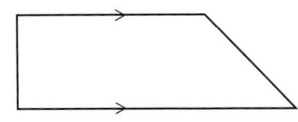

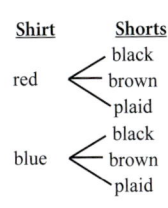

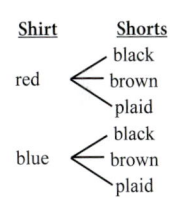

Two-Way Frequency Table	A table that shows how many times a value occurs for a pair of categorical data.	Tabla de Frecuencia Bidireccional	Una tabla que muestra cuántas veces aparece un valor de un par de datos categóricos.

U-V-W

Unit Rate	A rate with a denominator of 1.	Índice de Unidad	Un índice con un denominador de 1.
Univariate Data	Data that describes one variable (i.e., scores on a test).	Data Univariados	Datos que describen una variable (es decir: puntajes en una prueba).
Variable	A symbol that represents one or more numbers.	Variable	Un símbolo que representa uno o más números.
Vertex	The minimum or maximum point on a parabola.	Vértice	El mínimo o máximo punto en una parábola.
Vertex of a Solid	The point where three or more edges meet.	Vértice de un Sólido	El punto donde tres o más bordes se encuentran.
Vertex of a Triangle	A point where two sides of a triangle meet.	Vértice de un Triángulo	Un punto donde dos lados de un triángulo se encuentran.

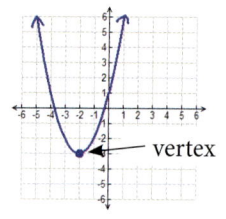

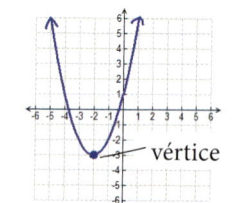

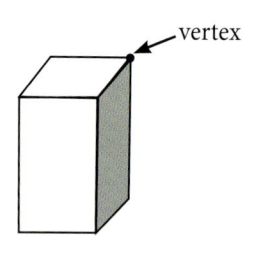

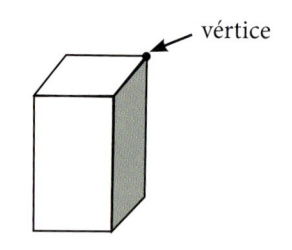

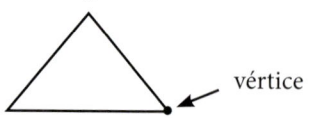

Glossary ~ Glosario

Vertex of an Angle	The common endpoint of the two rays that form an angle.	Vértice de un Ángulo	El punto final en común de los dos rayos que forma un ángulo.
Vertex Form	A quadratic function is in vertex form when written $f(x) = a(x - h)^2 + k$ where $a \neq 0$.	Forma De Vértice	Una función cuadrática es en forma general cuándo escrito $f(x) = a(x - h)^2 + k$ donde $a \neq 0$.
Vertical Angles	Non-adjacent angles with a common vertex formed by two intersecting lines.	Ángulos Verticales	Ángulos no adyacentes con un vértice en común formado por dos rectas intersecantes.
Vertical Line Test	A test used to determine if a graph represents a function by checking to see if a vertical line passes through no more than one point of the graph of a relation.	Examen Vertical De Línia	Un examen para determinar si una gráfica representa una función. Es utilizada para ver si una línia vertical que pasa a través de no más de un punto de la gráfica de una relación.
Volume	The number of cubic units needed to fill a three-dimensional figure.	Volumen	La cantidad de unidades cúbicas necesitadas para llenar un sólido.

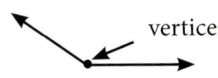

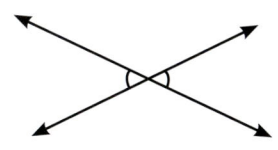

X-Y-Z

x-Axis	The horizontal number line on a coordinate plane.	Eje-x, Eje de la x	La recta numérica horizontal en un plano de coordenadas.

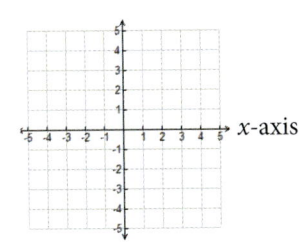

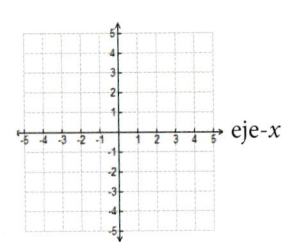

Glossary ~ Glosario **201**

y-Axis	The vertical number line on a coordinate plane.	Eje-*y*, Eje de la *y*	La recta numérica vertical en un plano de coordenadas.
y-Intercept	The point where a graph intersects the *y*-axis.	Intersección *y*	El punto donde una gráfica interseca el eje-*y*.
Zero Pair	One positive integer chip paired with one negative integer chip. 1 + (−1) = 0	Par Cero	Un chip entero positivo emparejado con un chip entero negativo. 1 + (−1) = 0
Zero Product Property	If a product of two factors is equal to zero, then one or both of the factors must be zero.	Propiedad De Producto Cero	Si un producto de dos factores es igual a cero, uno o ambos de los factores debe ser cero.
Zeros	The *x*-intercepts of a quadratic function. zeros, roots, *x*-intercepts	Ceros	Las intersecciones-*x* de una función cuadratica. ceros, raíces, intersecciones-*x*

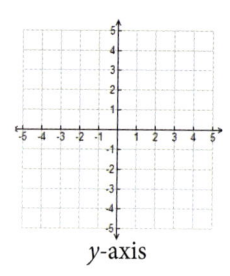

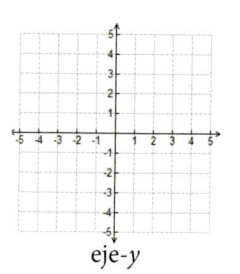

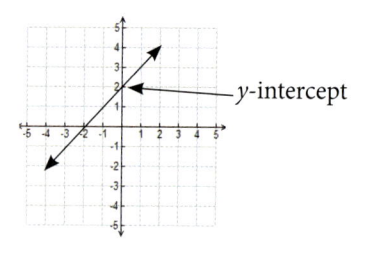

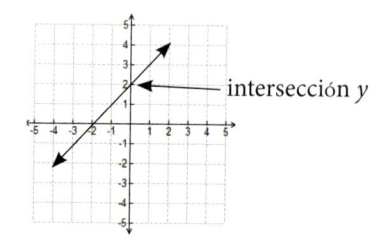

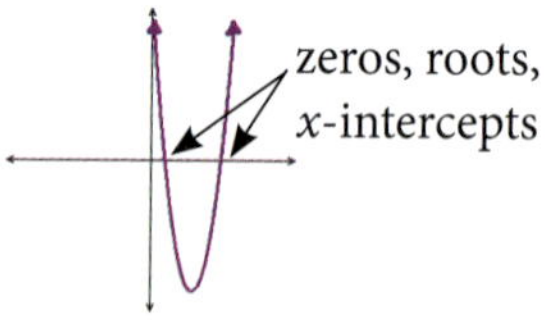

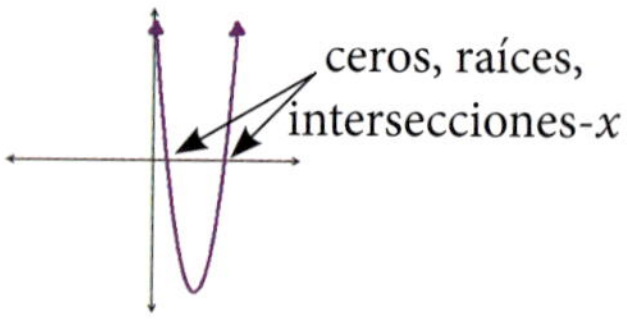

SELECTED ANSWERS

BLOCK 1

Lesson 1.1

1. Answers may vary (i.e., $\frac{1}{2}$, $\frac{4}{8}$). **3.** Answers may vary (i.e., 3 : 1, 12 : 4). **5.** Answers may vary (i.e., 1 to 1, 2 to 2). **7.** $\frac{2}{3}$, 2 : 3, 2 to 3 **9.** $\frac{4}{1}$, 4 : 1, 4 to 1 **11.** $\frac{1}{1}$ or 1, 1 : 1, 1 to 1 **13. a)** $\frac{7}{8}$ **b)** $\frac{7}{15}$ **c)** $\frac{8}{15}$ **15. a)** $\frac{1}{2}$ **b)** $\frac{1}{1}$ **17. a)** $\frac{1}{3}$ **b)** $\frac{1}{2}$ **19. a)** The team won 3 games for every 4 games it lost. **b)** 3 : 7 **21. a)** 5 : 8 **b)** 100 **c)** Comedy $\frac{1}{4}$; Drama $\frac{1}{5}$; Action $\frac{2}{5}$; Adventure $\frac{1}{10}$; Cartoon $\frac{1}{20}$ **23.** 12 squares **25. a)** 28; See student work. **b)** 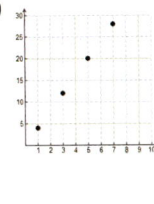
27. a) 12; See student work.
b) 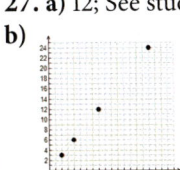
29. a) Answers may vary (e.g., 6 : 10 and 9 : 15) **b)** Kyle is incorrect because $\frac{12}{16} \neq \frac{3}{5}$
31.

	2	4	6	8
1 : 3	6	12	18	24
2 : 5	5	10	15	20

18, 20; the ratio 1 : 3 gives a larger second value because it multiplies the first value in the ratio by 3 and the ratio 2 : 5 multiplies the first value by $\frac{5}{2}$ which is less than 3.

Lesson 1.2

1. ratio: 5, terms: 625, 3125 **3.** ratio: $\frac{1}{2}$, terms: 50, 25 **5.** ratio: $\frac{1}{2}$, terms: $\frac{3}{16}$, $\frac{3}{32}$ **7. a)** 128, 64, 32, 16, 8, 4, 2, 1 **b)** $\frac{1}{2}$ **c)** No, the numbers become small fractions that get close to 0, but never equal 0 (and you cannot have a fraction of a person). **9.** 10000, 1000, 100, 10, 1, ... **11.** not geometric **13.** geometric → ratio: $\frac{1}{10}$ **15.** geometric → ratio: $\frac{1}{5}$ **17. a)** 1 **b)** 2 **c)** 4 **d)** 8, 16, 32, 64, ... (fold it 7 times) **e)** Yes, the number of rectangles always doubles which means the sequence has a ratio of 2. **19.** Answers may vary. **21.** Answers may vary (i.e., 2 :3, 12 : 18). **23.** Answers may vary (i.e., $\frac{2}{3}$, $\frac{4}{6}$). **25. a)** 5 : 7 **b)** 5 : 12 **c)** 7 : 12

Lesson 1.3

1. 4 inches **3.** 15 minutes **5.** 8 ounces **7.** 3 feet; See student work. **9.** 6 pints **11.** 8 quarts **13.** 2.5 tons **15.** 42 feet **17.** 2.5 feet **19.** 300 minutes **21.** 17,400 inches **23.** $\frac{1}{2}$, 1 : 2, 1 to 2 **25.** $\frac{3}{4}$, 3 : 4, 3 to 4 **27.** ratio: $\frac{1}{2}$, terms: $\frac{1}{2}$, $\frac{1}{4}$ **29.** ratio: 6, terms 216, 1,296

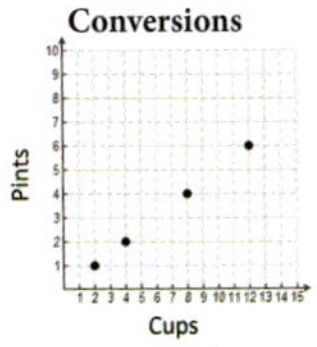

Lesson 1.4

1. 13 centimeters **3.** 1 meter **5.** Yes, there are 1,000 meters in every kilometer. **7.** 7 grams; See student work. **9.** 50 milliliters **11.** 600 centimeters **13.** 5 kilometers **15.** 3,000 milliliters **17.** 7,000 meters **19.** 92 kilograms **21.** 18,000 *cm* **23.** Answers may vary. **25.** 6,000 pounds **27.** ratio: 10, terms: 100000, 1000000

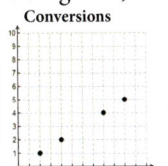

Lesson 1.5

1. a) 32 *ft* **b)** 384 *in* **3. a)** $2\frac{1}{2}$ yd **b)** $7\frac{1}{2}$ ft **5. a)** 64 square feet **b)** 9,216 square inches **7. a)** $\frac{3}{8}$ square yards **b)** $3\frac{3}{8}$ square feet **9.** 10,000 square centimeters; See student work. **11. a)** 180 square feet **b)** 20 square yards **c)** $839 **13. a)** 174 *ft*; See student work. **b)** $2,088 **15. a)** small: 24 *m*; large: 36 *m* **b)** 2 : 3 **c)** small: 24 square meters and large: 54 square meters **d)** 4 : 9 **17.** ratio $\frac{1}{2}$, terms: 1.875, 0.9375 **19.** ratio: 2, terms: 80, 160

Block 1 Review

1. $\frac{1}{2}$, 1 : 2, 1 to 2 **3.** $\frac{2}{1}$, 2 : 1 or 2 to 1 **5.** 16 squares **7. a)** 12; See student work. **b)** 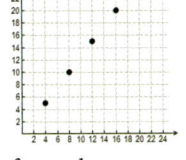
9. a) 7 : 9 **b)** 7 : 16 **c)** 9 : 16 **11. a)** 8 : 15 **b)** 2 : 3 **13. a)** 3 people **b)** 9 people **c)** 27 people **d)** 1, 3, 9, 27; Yes, it is geometric because three people will always smile for each person which means there is a ratio of 3 between the terms. **15.** ratio $\frac{1}{3}$, terms: $\frac{1}{81}$, $\frac{1}{243}$ **17.** 3, 6, 12, 24, 48, ... **19.** 10, 30, 90, 270, 810, ... **21.** not geometric **23. a)** 2,425 feet **b)** 600,000 square centimeters **25.** 2 cups **27.** 56 days **29.** 5.5 pints **31.** 10,560 feet **33.** 4 yards **35.** 365 days, 12 months; See student work. **37.** 28 centimeters **39.** 60,000 millimeters, 0.06 kilometers, 600 decimeters; See student work. **41.** 70 decimeters **43.** 0.8 kilometers **45.** 2,000 liters **47.** 100,000 centimeters **49. a)** $2\frac{1}{3}$ yd **b)** 7 ft **51. a)** 2 square yards **b)** 18 square feet **53. a)** 60 square meters **b)** 600,000 square centimeters **55.** Answers may vary but must have a product of 54 square feet.

BLOCK 2

Lesson 2.1

1. 0.4 **3.** 0.7 **5.** 0.625 **7.** 2.5 **9.** 10.6 **11.** Natalia ~ $\frac{5}{8}$ = 5 ÷ 8; Tricia divided the denominator by the numerator
13. $\frac{1}{4}$; 0.1875< 0.25 **15.** $\frac{3}{10}$ **17.** $\frac{1}{2}$ **19.** $\frac{3}{20}$ **21.** $9\frac{3}{8}$ **23.** $4\frac{1}{50}$
25. $\frac{1}{4}$ = 0.25
27. a)

Decimal	0.1	0.2	0.3	0.4	0.5	0.6	0.7	0.8	0.9	1
Fraction	$\frac{1}{10}$	$\frac{1}{5}$	$\frac{3}{10}$	$\frac{2}{5}$	$\frac{1}{2}$	$\frac{3}{5}$	$\frac{7}{10}$	$\frac{4}{5}$	$\frac{9}{10}$	1

b) Answers may vary **c)** 7.8 cm **29.** Siri; See student work.
31. ratio: 5, terms: 2500, 12500 **33.** 30 feet **35.** 2 kilometers
37. 2,500 meters

Lesson 2.2

1. $0.\overline{6}$ **3.** $0.\overline{7}$ **5.** 0.75 **7.** $0.\overline{09}$ **9.** $0.\overline{4}$ **11. a)** $\frac{1}{3} + \frac{2}{3} = 1$
b) $0.\overline{3} + 0.\overline{6} = 0.\overline{9}$ **c)** The answers look different, but should be and are the same. $1 = 0.\overline{9}$ **13.** 0.3 **15.** 10.3 **17.** 5.07
19. 11.21 **21.** 23.444 **23.** 0.286 **25.** 0.78 **27.** 0.83
29. a) Marci **b)** No, both round to 12.03 seconds
31. 3.24 minutes because that rounds to 3.2, but 3.25 minutes would round to 3.3 minutes **33.** $\frac{3}{4}$ **35.** $\frac{1}{3}$; $0.\overline{3} > 0.3$
37. $\frac{1}{2}$; 0.4 < 0.5

Lesson 2.3

1. $\frac{30 \text{ miles}}{1 \text{ hour}}$ **3.** $\frac{20 \text{ miles}}{1 \text{ gallon}}$ **5.** $\frac{\$1.10}{1 \text{ pen}}$ **7.** $\frac{4 \text{ pounds}}{1 \text{ inch}}$ **9.** $\frac{3 \text{ kilometers}}{1 \text{ hour}}$
11. $9.00 per ticket; See student work. **13.** 23.5 miles per gallon; See student work. **15.** Music Hooray is cheaper because Songs-R-Us charges $0.50 per song. **17.** $0.\overline{3}$
19. 0.375 **21.** $\frac{1}{4}$ **23.** $1\frac{3}{10}$ **25.** 0.3 **27.** 5.8 **29.** 21 girls; See student work.

Lesson 2.4

1. $30.00 **3.** 420 words **5.** 8 gallons **7.** 240 miles **9.** $180
11. $\frac{\$0.40}{1 \text{ pencil}}$ **13.** $\frac{0.25 \text{ meters}}{1 \text{ second}}$ **15.** $\frac{\$0.67}{1 \text{ book}}$ **17.** $4.84 **19.** Aaron could use equivalent fractions and multiply the numerator and denominator by 3: $\frac{8 \text{ miles}}{2 \text{ hours}} = \frac{24 \text{ miles}}{6 \text{ hours}}$; 24 miles **21.** No, she has $1.75 left and the cookie and popcorn cost $2.00; see Student work. **23.** 3.5 centimeters **25.** 15 feet **27.** 4.5 miles per hour

Lesson 2.5

1. $\frac{\$6.75}{3 \text{ sandwiches}}$ **3.** $\frac{162 \text{ miles}}{6 \text{ gallons}}$ **5.** Christine; See student work.
7. package of 6 pens **9. a)** Jamal because he drove 55 miles per hour and Mark drove 50 miles per hour. **b)** Mark drove 300 miles. Jamal drove 330 miles. **11.** $40.00
13. a) 3 packages **b)** $60.00 **c)** $40.00 **d)** 4 packages of 15 notebooks **e)** 20 pkg → $1.00 per notebook; 15 pkg → $0.67 per notebook **15.** $16.00 because he will pay $0.80 per balloon in the 5-balloon package instead of $0.90 per balloon in the 4-balloon package. Either way, he will not have balloons left over or buy extra balloons. **17.** $0.\overline{6}$ **19.** 1.5
21. a) 2 : 3 **b)** 2 : 1

Lesson 2.6

1.

Hours	1	2	3	4	5	6
Total Miles Traveled	6	12	18	24	30	36

3. 2.25 miles; See student work. **5.** 1.5 hours; See student work.
7. 20 miles; see student work **9.** $\frac{9 \text{ miles}}{1 \text{ hour}} > \frac{8 \text{ miles}}{1 \text{ hour}}$; $\frac{9 \text{ miles}}{1 \text{ hour}}$
11. $\frac{7 \text{ yards}}{1 \text{ day}} > \frac{6 \text{ yards}}{1 \text{ day}}$; $\frac{14 \text{ yards}}{2 \text{ days}}$ **13.** Hillary; See student work.
15. Red crew; see student work **17.** 12 **19.** 100 **21.** B **23.** B **25. a)** 36,960 feet per hour **b)** See student work: $\frac{36,960 \text{ ft}}{1 \text{ hour}} \cdot \frac{1 \text{ hour}}{60 \text{ min}} = \frac{616 \text{ ft}}{1 \text{ min}}$ **27.** 9,000 **29.** 300 **31.** 3.25 **33.** 0.125 **35.** $3\frac{1}{3}$
37. $1\frac{3}{5}$

Block 2 Review

1. 0.2 **3.** 2.3 **5.** $\frac{3}{12}$; 0.25 > 0.2 **7.** $\frac{3}{5}$ **9.** $\frac{1}{8}$ **11.** $0.\overline{6}$ **13.** $0.\overline{5}$
15. 0.5 **17.** 0.2 **19.** 0.7 **21.** 0.13 **23.** Lonnie: 5.64, Porter: 5.55; See student work. **25.** $\frac{30 \text{ miles}}{1 \text{ gallon}}$ **27.** $\frac{10 \text{ miles}}{1 \text{ hour}}$
29. $\frac{\$0.50}{1 \text{ picture}}$ **31.** $4.00 per notebook **33.** 25.5 miles per gallon; see student work **35.** Lia found miles per gallon:
$\frac{105 \text{ miles}}{5 \text{ gallons}} = \frac{21 \text{ miles}}{1 \text{ gallon}}$. Her dad found gallons per mile:
$\frac{5 \text{ gallons}}{105 \text{ miles}} \approx \frac{0.0476 \text{ gallons}}{1 \text{ mile}}$ which is about 0.05 gallons per mile.
37. 400 **39.** $28.00 **41.** 330 miles; See student work.
43. $3.33 per toy car **45.** 22 miles per gallon **47.** $11.25; See student work. **49.** $\frac{42 \text{ miles}}{4 \text{ hours}}$ **51.** Casey; See student work.
53. $18.00 **55. a)** $48.00 **b)** 3 packages **c)** 3 packages of 8 pens because packages of 6 pens cost $2.00 per pen and packages of 8 pens cost $1.75 per pen. **57.** 0.7 miles
59. The frog that hops 1 meter every 4 minutes would win; See student work. **61.** 3 **63.** B **65. a)** 26,400 feet per hour
b) See student work. $\frac{26,400 \text{ ft}}{1 \text{ hour}} \cdot \frac{1 \text{ hour}}{60 \text{ min}} = \frac{440 \text{ ft}}{1 \text{ min}}$ **67.** 8,000

BLOCK 3

Lesson 3.1

1. $\frac{40}{100}$; 40% **3.** $\frac{10}{100}$; 10% **5.** $\frac{1}{20}$ **7.** $\frac{3}{4}$ **9.** $\frac{9}{20}$ **11.** $\frac{8}{25}$ **13.** $\frac{1}{250}$
15. 0.4 **17.** 0.82 **19.** 0.5 **21.** 0.005 **23. a)** ●○ **b)** ●●
c) 2 **d)** 200% **25.** 85% **27. a)** $\frac{7}{1000}$ **b)** 0.007 **c)** 7 out of 1,000 stores have it in stock **29. a)** Answers may vary. 30 squares → bed; 24 squares → dresser; 20 squares → desk **b)** 26% left over
31. 8,000 pounds **33.** 0.25 kilometers

Lesson 3.2

1. $\frac{1}{4}$, 0.25, 25% **3.** $\frac{2}{3}$, $0.\overline{6}$, $66\frac{2}{3}$% **5.** $\frac{3}{4}$, 0.75, 75% **7.** 4%
9. 56.5% **11.** 225% **13.** 50% **15.** $66\frac{2}{3}$% or $66.\overline{6}$% **17.** 125%
19. a) shade 3 squares **b)** 25% **c)** 75%
21. 0.75, $\frac{3}{4}$, 75%, $\frac{75}{100}$ **23.** 0.4 = $\frac{4}{10}$ → $\frac{4}{10}$ = $\frac{40}{100}$ → $\frac{40}{100}$ = 40%
25. 60%, $\frac{2}{3}$, 0.7 **27. a)** 45% **b)** 55% **29.** 40%; See student work. **31.** $\frac{1}{4}$ **33.** $\frac{9}{20}$ **35.** 0.33 **37.** 2.8 **39.** 24 red marbles; See student work.

204 *Selected Answers*

Lesson 3.3

1. 7 **3.** 3 **5.** 30 **7.** 44 **9.** 11.2 **11.** $0.45 **13.** 0.75 m or 75 cm
15. 56%; 1 m = 100 cm and 56 cm out of 100 cm is $\frac{56}{100}$ = 56%
17. 10, 20 **19.** 20, 30 **21.** line → 3.2 cm **23.** line → 7 cm
25. No, 30% of 60 is 18 because 0.3 × 60 = 18. **27.** 135%
29. 62% **31.** 25%

Lesson 3.4

1. 35 pansies **3.** 2.4 feet **5. a)** ≈ $8.99 **b)** $6.00 **7.** $169.15; See student work. **9.** $2.25 **11.** $6.74 **13.** $52.80; See student work. **15. a)** $1,440 **b)** $25,440 **17.** No, because a 50% discount followed by a 25% discount means the DVD will cost $7.50, while a 75% discount results in a cost of $5.00. See student work. **19.** $\frac{1}{10}$; 0.1 **21.** $1\frac{1}{10}$; 1.1 **23.** $\frac{41}{50}$; 0.82
25. 14 days per job **27.** Yan, because he biked 10.5 miles per hour while Eliza biked 10 miles per hour.

Lesson 3.5

1. certain **3.** equally likely **5.** equally likely **7.** likely → $\frac{5}{8}$ of the spinner is blue **9.** certain → these colors are 100% of the spinner **11.** $\frac{1}{4}$; 0.25 **13.** Answers may vary (describe something that is certain to happen). **15. a)** $\frac{7}{10}$; 0.7 **b)** It has mostly red jelly beans. **17. a)** 90%; $\frac{9}{10}$ **b)** Yes, 90% is very likely to occur **19. a)** $0.\overline{6}$%; $66\frac{2}{3}$% **b)** No, although the chance is high, it is not certain. **21.** 0.1, $\frac{1}{8}$, $\frac{3}{10}$, 15% **23.** chocolate (8 students), strawberry (12 students), vanilla (20 students)
25. 30%, $\frac{1}{3}$, 0.35 **27.** 0.7, 75%, $\frac{4}{5}$ **29.** 36 **31.** 12.5 **33.** 0.1
35. $4.34

Lesson 3.6

1. {heads, tails} **3.** {blue, blue, red} **5.** {blue, green, green, green, green, red, red} **7.** $\frac{2}{5}$ or 0.4 or 40%
9. $\frac{1}{3}$ or $0.\overline{3}$ or $33.\overline{3}$% **11.** Yes, because 2 of the 10 responses were no and $\frac{2}{10} = \frac{1}{5}$. **13. a)** truck → $\frac{2}{15}$; van → $\frac{4}{15}$; SUV → $\frac{1}{5}$; car → $\frac{2}{5}$ **b)** car **15. a)** 40 **b)** $\frac{1}{5}$, 0.2, 20% **c)** $\frac{3}{8}$, 0.375, 37.5%
d) $\frac{7}{40}$, 0.175, 17.5% **e)** $\frac{1}{4}$, 0.25, 25% **f)** hearts **17.** $\frac{2}{11}$, 0.2, 23%
19. $324; See student work.

Lesson 3.7

1. $\frac{1}{6}$ **3.** $\frac{1}{2}$ **5.** $\frac{5}{6}$ **7.** $\frac{1}{11}$ **9.** $\frac{4}{11}$ **11.** $\frac{4}{11}$ **13.** $\frac{1}{20}$ **15.** $\frac{1}{2}$ **17.** $\frac{1}{2}$, 0.5, 50%
19. $\frac{7}{8}$, 0.875, 87.5% **21.** Answers may vary (e.g., P(yellow) = P(green) = $\frac{1}{8}$). **23.** $\frac{4}{7}$ **25. a)** $\frac{1}{9}$ **b)** $\frac{1}{3}$
27. (coloring varies) green → 6 sectors; blue → 2 sectors; red → 1 sector; yellow → 3 sectors **29. a)** $\frac{1}{4}$ **b)** $\frac{3}{5}$ **31.** 320
33. 4.5

Lesson 3.8

1. perimeter = 22 in; area = 28 square inches; see student work **3.** perimeter = 24 m; area = 24 square meters; see student work **5.** perimeter = 40 yd; area = 100 square yards; See student work. **7.** $\frac{1}{3}$ **9.** $\frac{1}{2}$ **11.** $\frac{8}{25}$ **13.** $\frac{3}{14}$ **15.** $\frac{1}{8}$
17. 245 miles **19.** $4.28 **21.** 3.6 **23.** 6 **25.** 9

Block 3 Review

1. $\frac{70}{100}$; 70% **3.** $\frac{20}{100}$; 20% **5.** $\frac{1}{4}$ **7.** 0.35 **9.** 0.01 **11.** If a game is 20% off the original price, then you pay 100% − 20% = 80% of the original price. They are both correct. **13.** $\frac{3}{4}$; 0.75, 75%
15. $\frac{3}{5}$; 0.6; 60% **17.** 80% **19.** 25% **21.** $66\frac{2}{3}$% **23.** 30%, $\frac{1}{3}$, 0.34
25. No, because 1.5 × 100 = 150 so 1.5 is equivalent to 150%.
27. 20 **29.** 100 **31.** 120 **33.** $0.45 **35.** 0.75 m or 75 cm
37. line is 8.8 cm **39.** $675; See student work. **41.** $55.20; See student work. **43.** $378; See student work. **45.** equally likely **47.** impossible **49. a)** $\frac{3}{4}$; 0.75 **b)** mostly pink jelly beans **51. a)** football; See student work. **b)** football → 22 people; baseball → 6 people; basketball → 12 people
53. {1, 2, 3, 4, 5, 6} **55.** $\frac{7}{20}$ **57. a)** $\frac{2}{5}$ or 0.4 or 40% **b)** $\frac{6}{25}$ or 0.24 or 24% **c)** $\frac{9}{25}$ or 0.36 or 36% **d)** $\frac{16}{25}$ or 0.64 or 64% **59.** No, the number 2 is one of six numbers on the number cube so P(2) = $\frac{1}{6}$. **61.** $\frac{1}{3}$ **63.** $\frac{1}{6}$ **65.** $\frac{1}{4}$ **67.** $\frac{5}{8}$ **69.** $\frac{364}{365}$; See student work.
71. perimeter = 80 in; area = 400 square inches; See student work. **73.** perimeter = 36 cm; area = 54 square centimeters; See student work. **75.** $\frac{1}{4}$ **77.** $\frac{2}{5}$

BLOCK 4

Lesson 4.1

1. Answers may vary. Numerical data uses numbers and categorical data uses words. **3.** categorical **5.** numerical
7. categorical **9.** A (explanations vary). Question B would not give variability in answers. **11.** A (explanations vary). There are more possibilities for the number of hours worked.
13. Answers may vary. People going to a pizza restaurant are more likely to say that pizza is their favorite. **15.** Answers may vary. Answering the phone is optional so it may not be representative of all opinions. **17.** Answers may vary. Responding is optional.

19. a)

[Gas Prices line graph showing Cost per Gallon from January-09 to January-13, ranging from about 1.8 to 3.6]

b) The original graph (explanations vary). It appears that gas prices aren't changing as much. **c)** The graph from part a (explanations vary). It appears that gas prices are increasing sharply. **21.** 0.4 **23.** $0.\overline{3}$ **25.** 0.33 **27.** 0.27

Lesson 4.2

1. a) He did not subtotal the sum of values before dividing. **b)** (12 + 4 + 8 + 11 + 10) ÷ 5 **c)** 9 **3.** Elena is correct. No number appears more often than any other number. There is no mode. **5.** 12 **7.** 14 **9.** 16 **11.** 8, 10 **13.** 13 **15.** 43 **17.** Mean = 10.5; Median = 9; Mode = 5, 10 **19.** Mean ≈ 16.9; Median = 20; Mode = 12, 20, 22 **21.** Mean ≈ 9.1; Median = 9; no Mode **23.** 27 **25.** 98 **27.** $35; See student work. **29.** 20; 52; See student work. **31.** Answers may vary; half of the people in her city are over 28 years old and half are under 28 years old. There are some people who are quite a bit older because the mean is higher than the median. **33.** $0.\overline{6}$ **35.** B (explanations vary). Question B gives more variability in answers. **37.** Answers may vary.

Lesson 4.3

1. a) [dot plot from 1 to 10]

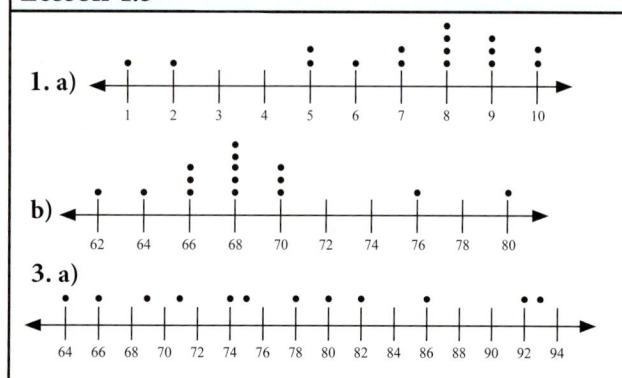

b) [dot plot from 62 to 80]

3. a) [dot plot from 64 to 94]

b) Answers may vary. Since there is no stacking of the dots, a dot plot may not be the best way to display this data.

5. a) [dot plot from 0 to 6] **b)** 1 **c)** 22 **d)** 39 **e)** ≈ 1.77

7. Answers may vary. The heights of basketball players would be greater, in general. Also, they might be more varied (larger range) than the heights of twelve average males.

9. a) mode **b)** List the numbers for each value in the dot plot from least to greatest and find the one value in the middle of the data list or the average of the two values in the middle of the data list.

11. Answers may vary. [dot plot from 50 to 80]

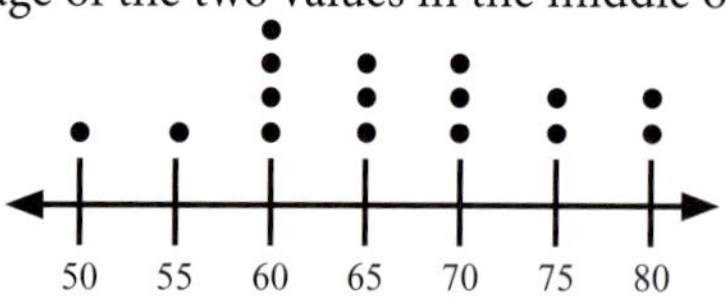

13. Answers may vary.
15. mean = 25; median = 24; mode = 20, 32
17. 79 **19.** categorical

Lesson 4.4

1. A bar graph displays data in categories. A histogram displays data in the form of numbers. **3. a)** 12 **b)** 1.5 pounds **c)** Eliza. The interval starts with the number on the left and goes up to (but does not include) the number on the right. **d)** 10-11.5 **5.** Intervals vary.

Interval	Frequency
1940-1950	2
1950-1960	4
1960-1970	8
1970-1980	1

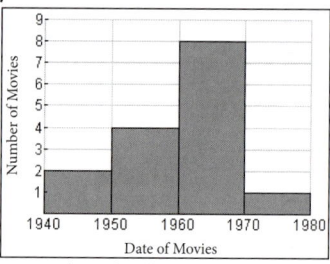

[histogram of Date of Movies vs Number of Movies]

7. a) 15 **b)** 2 **c)** 1 **d)** 6-8 hours **e)** 7 **9. a)** 0-10 years **b)** 440 **c)** ≈ 52%; see student work **11. a)** Answers may vary. There would be too few intervals. **b)** Answers may vary. Asking students their age in months would give more variability. **13.** No, data in a histogram is displayed in intervals by frequency, not by specific numbers that are needed to calculate mean, median and mode. **15.** 23 **17.** 50 seconds; See student work.

Lesson 4.5

1. minimum, 1st quartile, median, 3rd quartile, maximum
3. a) She didn't scale the horizontal axis evenly.
b) [box plot 6 to 22]
5. 7 ~ 14 ~ 19 ~ 25 ~ 27
7. 52 ~ 64.5 ~ 77 ~ 88 ~ 99
9. 6
11. 14

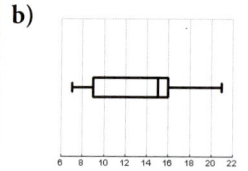

13. [box plot 4 to 32] **15.** [box plot 48 to 72]

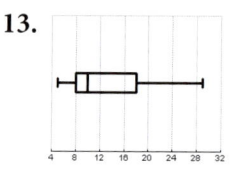

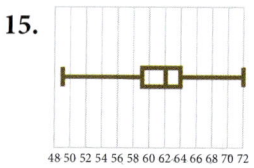

17. 3 ~ 9 ~ 14 ~ 18 ~ 20 [box plot 2 to 20]

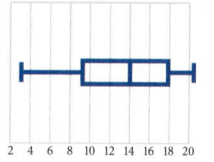

19. 3 ~ 10 ~ 15 ~ 18 ~ 35 [box plot: Cost of Kristen's Stuffed Animals]

21. a) [histogram] **b)** 15-18 **c)** Answers may vary. 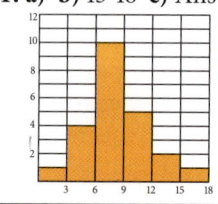 **d)** Dot plot; a histogram does not show individual values. **e)** Answers may vary.

Selected Answers

Lesson 4.6

1. a) mean = 36; median = 39; mode = 37 **b)** Yes; 8 is an outlier. **c)** median **3. a)** They are the same ($124).
b) Tamiqua's median is $3 higher. **c)** Tamiqua; she earned more than $120 almost every week. **d)** Jim; Jim earned more than $130 twice, but Tamiqua never earned that much.
5. a) Answers may vary. Travis had the higher median score.
b) Answers may vary. Alden had the higher mean score.
c) Answers may vary. **7.** Dot Plot, List of Values

9. a) 2 **b)** [number line dot plot]
c) 9; 9; They add up to the same value.
11. 18~29~37~46~50
13. Asking your friends does not give a random sample. [box plot]

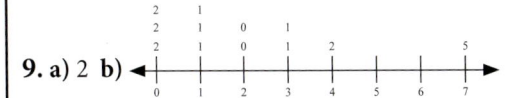

Lesson 4.7

1. The mean absolute deviation is the average distance each number is away from the mean of a data set. **3.** −7, −2, 1, 3, 5
5. a) 8 **b)** −6, −3, 0, 1, 2, 2, 4 **c)** 6, 3, 0, 1, 2, 2, 4 **d)** ≈ 2.57
7. a) 260; Yes **b)** $22.\overline{2}$; No **9.** Answers may vary. One possible solution: 1, 10, 10, 11, 11, 12, 22 (mean absolute deviation ≈ 3.43; IQR = 2 **11. a)** minimum = 8; maximum = 32; Answers may vary; reasonable interval width is 3. **b)**

Hours Worked	Tally
6-9	1
9-12	0
12-15	0
15-18	0
18-21	4
21-24	3
24-27	5
27-30	0
30-33	1

c) [histogram]

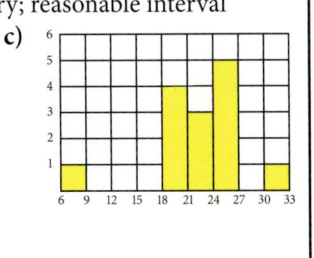

13. The mode; (explanations may vary). The value of 8 shows up far more often than the other values.

Block 4 Review

1. numerical **3.** numerical **5.** B (explanations vary); the second question gives more variability in the answers.
7. People outside a pizza restaurant are more likely to eat pizza frequently; instead, take a random sample of people outside a grocery store. **9.** The vertical axis does not start at 0.

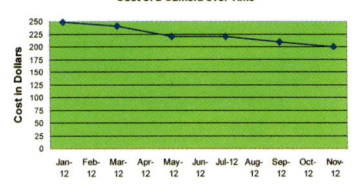

[Cost of a Camera over Time graph]

11. mean = 17; median = 16; mode = 14, 23
13. mean = 40.7; median = 40.5; mode = 38 **15.** 16
17. 6, 15; See student work. **19.** 12 **21.** 6.5
23. a) Length of Students' Shot Put Throws (feet) [dot plot]

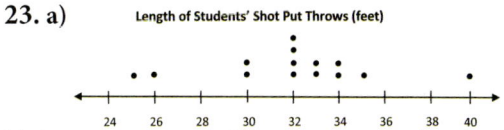

b) Answers may vary. There are many values in the middle, clustered from 30-35 feet. **25.** 4, 12 **27. a)** minimum = $1; maximum = $2.15; answers may vary; a reasonable interval width might be $0.20; See student work.

b)

Interval	Frequency
1.00-1.20	4
1.20-1.40	2
1.40-1.60	5
1.60-1.80	2
1.80-2.00	0
2.00-2.20	1

c) [histogram]

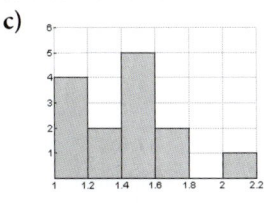

29. 25 **31.** 4~7.5~15~19.5~25 **33. a)** [box plot]
b) IQR = 9 **35. a)** mean ≈ 66.9; median = 66; mode = 65, 67
b) Yes, answers may vary; 76 is quite a bit larger than the rest of the values.
c) Answers may vary. There are two modes and the outlier affects the mean. The median would probably best represent the data. **37.** Answers may vary. The median. There is no mode and the mean would be affected by the outlier, 30.
39. a) 12 **b)** 1 **c)** 2 **d)** 8; 8; These values are the same.
41. 7, 1, 1, 2, 5; MAD = 3.2; See student work. **43. a)** 32 miles
b) 2, 2, 12, 12, 2, 2 **c)** $5.\overline{3}$ **d)** On average, Chris' bike rides are about 5.3 miles away from the mean.

Selected Answers **207**

INDEX

1st Quartile (Q1)
    in five-number summary, 141

3rd Quartile (Q3)
    in five-number summary, 141

A
Absolute deviation, 148

Area, 22
    rectangles, 23
    triangles, 23

B
Bias, 118
    in data collection, 118
    misleading graphs, 119

Box-and-whisker plot, 143

C
Career Focus
    ASL Interpreter, 166
    Coach, 70
    Commercial Truck Driver, 114
    Nurse, 32

Categorical data, 117
    Explore! A Question of Statistics, 117

Centimeter, 18

Compare rates, 54

Complements (probability), 101

Conversions
    area, 22
    customary units, 14
    decimals to fractions, 26
    decimals to percents, 78
    Explore! The Patio, 22
    fractions to decimals, 36
    metric units, 18
    percents to decimals, 74
    percents to fractions, 74
    perimeter, 22
    rate, 61

Cup, 14

Customary system, 14
    conversions between units, 14
    units of length, 14
        foot, 14
        inch, 14
        mile, 14

Customary system (cont)
    units of length
        yard, 14
    units of weight, 14
        ounce, 14
        pound, 14
        ton, 14
    units of volume, 14
        cup, 14
        fluid ounce, 14
        gallon, 14
        pint, 14
        quart, 14

D
Day, 14

Decimals
    converting percents to decimals, 74
    converting to fractions, 36
    converting to percents, 79
    repeating, 39
    rounding, 39
    writing percents as decimals, 74

Discount (percents), 85

Dot plot, 128
    Explore! How Tall?, 129

E
Explore!
    A Question of Statistics, 117
    At the Restaurant, 86
    Back and Forth, 35
    Calculators and Fractions, 40
    Compare Students, 3
    Counting Pets, 122
    How Tall?, 129
    Kieran's Room, 77
    Match the Rates, 48
    Mercury's Rising, 154
    Number Patterns, 10
    The Patio, 22
    Percents, 73
    Rolling a 3, 96
    Shopping Sales, 54
    Sum of Two Number Cubes, 100
    What are My Chances of Winning?, 105
    What's the "Mean"ing?, 148

F
Five-number summary, 141
    Box-and-whisker plot for, 143

Fluid ounce, 14

Foot, 14

Fractions
    converting percents to fractions, 74
    converting to percents, 78
    converting to decimals, 36, 39
    as repeating decimals, 39
    as terminating decimals, 39
        Explore! Back and Forth, 35
    equivalent fractions, 48

Frequency Table, 135

G
Gallon, 14

Geometric probability, 106
    Explore! What are My Chances of Winning?, 105

Geometric sequence, 9
    Explore! Number Patterns, 10
    ratio, 9
    term, 9

Gram, 18

H
Histogram, 134
    frequency table for, 135
    interval width of, 135-136
    making a, 135

Hour, 14

I
Inch, 14

Interquartile range (IQR), 141

J

K
Kilogram, 18

Kilometer, 18

L
Length
    customary system, 14
        foot, 14
        inch, 14
        mile, 14
        yard, 14

    metric system, 18
        centimeter, 18

metric system (cont)
   kilometer, 18
   meter, 18
   millimeter, 18

M

Mass (metric units), 18
   gram, 18
   kilogram, 18
   milligram, 18

Maximum, 124
   in five-number summary, 141

Mean, 122
   absolute deviation from the, 148
   calculator note, 123
   Explore! What's the "Mean"ing?, 148
   outliers affect on, 149

Measurement
   customary, 14
   metric, 18

Measures of center, 122
   Explore! Counting Pets, 122
   mean, 122
   median, 122
   mode, 122
   which best represents a data set, 149

Median, 122
   In five-number summary, 141

Meter, 18

Metric system, 18
   conversions between units, 18
   units of length, 18
      centimeter, 18
      kilometer, 18
      meter, 18
      millimeter, 18
   units of mass, 18
      gram, 18
      kilogram, 18
      milligram, 18
   units of volume, 18
      liter, 18
      millimeter, 18

Mile, 14

Milligram, 18

Milliliter, 18

Millimeter, 18

Minimum, 124
   in five-number summary, 141

Minute, 14

Misleading graph, 119

Month, 14

Motion rates, 60

Mode, 122

N

Numerical data, 117
   Explore! A Question of Statistics, 117

O

Ounce, 14

Outcomes, 94

Outlier, 149

P

Percent
   applications, 85
   discount, 85
   Explore! Kieran's Room, 77
   Explore! Percents, 73
   of a number, 81
   sales tax, 86
   tip, 86
      Explore! At the Restaurant, 86
   write decimals as percents, 78
   write fractions as percents, 78
   writing as fractions, 74
   writing as decimals, 74

Perimeter
   change in units, 20

Pint, 14

Pound, 14

Probability, 90
   complements, 91
   geometric, 106
   experimental, 95
   introduction, 90
   outcomes, 94
   sample space, 94
   theoretical, 100
   trial, 94
   written as decimal, fraction, percent, 91

Q

Quart, 14

Quartiles, 141
   in box-and-whisker plot, 143

R

Range, 124
   versus interquartile range, 143

Rate, 44
   conversion, 61
   Explore! Match the Rates, 48
   Explore! Shopping Sales, 54
   motion rates, 60
   problem solving, 49
      using equivalent fractions, 49
      using unit rates, 49
   unit rate, 44, 49

Ratio, 3
   Explore! Compare Students, 3
   geometric sequence, 9
   writing, 4

Rectangle
   area formula, 23
   geometric probability, 106
   perimeter and area, 23

Repeating decimal, 39

Rounding decimals, 40

S

Sales tax, 86

Sample space, 94

Second (time), 14

Sequence (geometric), 9

Statistics, 117
   analyzing, 148-150
   Explore! What's the "Mean"ing?, 148
   bias of, 118
   Explore! A Question of Statistics, 117
   large data set versus small data set, 149
   outlier in, 149

T

Term, 9

Terminate, 39

Index **209**

Theoretical probability, 100
  equally likely, 91
  Explore! Sum of Two Number Cubes, 100

Time (units of), 14
  day, 14
  hour, 14
  minute, 14
  month, 14
  second, 14
  week, 14
  year, 14

Tip, 86
  Explore! At the Restaurant, 86

Ton, 14

Triangles
  area formula, 24
  geometric probability, 106
  perimeter and area, 23

Trial, 94

U
Unit rate, 44

Units, *see* customary system and metric system

V
Volume
  customary units of volume, 14
    cup, 14
    fluid ounce, 14
    gallon, 14
    pint, 14
    quart, 14
  metric units of volume, 18
    liter, 18
    millileter, 18

W
Week, 14

Weight
  customary units of weight, 14
    ounce, 14
    pound, 14
    ton, 14

X

Y
Yard, 14

Year, 14

Z

PROBLEM-SOLVING

UNDERSTAND THE SITUATION

- Read then re-read the problem.
- Identify what the problem is asking you to find.
- Locate the key information.

PLAN YOUR APPROACH

Choose a strategy to solve the problem:

- Guess, check and revise
- Use an equation
- Use a formula
- Draw a picture
- Draw a graph
- Make a table
- Make a chart
- Make a list
- Look for patterns
- Compute or simplify

SOLVE THE PROBLEM

- Use your strategy to solve the problem.
- Show all work.

ANSWER THE QUESTION

- State your answer in a complete sentence.
- Include the appropriate units.

STOP AND THINK

- Did you answer the question that was asked?
- Does your answer make sense?
- Does your answer have the correct units?
- Look back over your work and correct any mistakes.

DEFEND YOUR ANSWER

Show that your answer is correct by doing one of the following:

- Use a second strategy to get the same answer.
- Verify that your first calculations are accurate by repeating your process.

Problem-Solving **211**

SYMBOLS

Algebra and Number Operations

SYMBOL	MEANING
+	Plus or positive
−	Minus or negative
$5 \times n$, $5 \cdot n$, $5n$, $5(n)$	Times (multiplication)
$3 \div 4$, $4\overline{)3}$, $\frac{3}{4}$	Divided by (division)
=	Is equal to
≈	Is approximately
<	Is less than
>	Is greater than
%	Percent
$a : b$ or $\frac{a}{b}$	Ratio of a to b
$5.\overline{2}$	Repeating decimal (5.222…)
≥	Is greater than or equal to
≤	Is less than or equal to
x^n	The nth power of x
(a, b)	Ordered pair where a is the x-coordinate and b is the y-coordinate
±	Plus or minus
$\sqrt{x}$	Square root of x
≠	Not equal to
$x \stackrel{?}{=} y$	Is x equal to y?
$|x|$	Absolute value of x
$P(A)$	Probability of event A

Geometry and Measurement

SYMBOL	MEANING
≅	Is congruent to
~	Is similar to
∠	Angle
$m\angle$	Measure of angle
$\triangle ABC$	Triangle ABC
$\overline{AB}$	Line segment AB
$\overrightarrow{AB}$	Ray AB
AB	Length of AB
π	Pi (approximately $\frac{22}{7}$ or 3.14)
°	Degree

212 *Symbols*